# Unto These Skies We Came

**WORKS BY IRV BROUGHTON**

*Unto These Skies We Came*

*The Fires of Tangerine: Poems*

*Fielding the Cape Cod Past: Poems*

*Where the Wings Grow*

*Conversations with Pioneering Women Pilots*

*The Levees that Break in the Heart :* Novel

*The Gracious Afterward: Delta Stories*

*Gotcha! Watcha! Drugs are a Trap*

*The Lost Peninsula:*
*Adventures in Florida*

*The Long Net:*
*Poems about the Great Game of Basketball*

*ESPN: The Untold Story (with Stuart Evey)*

*Forever Remembered: The Fliers of World War II*

*Hangar Talk: Conversations with American Fliers*

*A Good Man: Fathers and Sons in Poetry and Prose*

*The Writer's Mind: Interviews with American Authors Volumes 1 - 3*

*Producers on Producing: The Making of Film and Television*

*The Art of the Interview in Television, Radio and Film*

*The Blessing of the Fleet: Poems*

**PLAYS**

*Smitten: The Internet Musical*

*Three to the Heart: A Life with Frank Stanford*

# Unto These Skies We Came

By Irv Broughton

Open Look Books
Spokane, Washington

*Unto These Skies We Came*





Open Look Books
Spokane, Washington

Book is available from Kindle, Amazon and fine Book Sellers.
Cover: Lilly Freya Ross

Interior Illustrations: Bonnie Shields
For Information or to schedule speaking engagements:

www.irvbroughton.com

## Acknowledgments

Some interviews appeared first in ***Where the Wings Grow*** Copyright 2019 by Irv Broughton. They include Gladys Buroker, Betty Pfister and Amelia Reid. Special thanks to Gene Colburn and Dennis Held.

*He used to talk about the air, and the huge way that it was . . . that it's alive all around you. You'd have to go up in an airplane to understand what he meant.*

James Dickey *Alnilam*

# Contents

# INTRODUCTION

This collection of interviews, *Unto These Skies We Came,* feels somewhat especially personal, since it's the planned final book of flier interviews, this one done over a period of close to fifty years. Am I that old? Best not fly to any conclusion. All told, I have interviewed in depth more than one-hundred fliers, from pioneers to military, all ground-breakers who have made aviation history in America. Many of the interviews were of everyday fliers, not famous, but persons who did their jobs and sacrificed greatly for their country. These men and women received no ticker-tape parades, no great accolades, but they need to be remembered. Such is the task of remembrance.

For me, flying has been a sort of obsession and will always remain so. Bad eyes kept me from ever pursuing it, but I've managed to enjoy and live wonderful adventures through my many interviews. After this book, I suppose I move on to other subjects and other pursuits, but I will always treasure the years spent in the company of some fascinating human beings.

Flying can test one and bring out the best, and the fliers represented here, I believe, to be the best. They are often brave beyond compare, brilliant beyond common understanding. They are modest and sometimes it took a coaxing to get them to talk. But once they did, I would on occasion feel like a child, nourished by the yarns that carry us, by which we learn about ourselves through others. Over these years, happily, the interest I held really derives from inspiration by these persons. I hope the reader feels likewise about these special people.

Irv Broughton

**For Kenzie, Sayer and Callie.**

# Albuquerque and the Birth of Ballooning

## Sidney Cutter

Sid Cutter is considered the "old man" in modern ballooning. A two-time national champion, as well as a promoter of the sport, it seems ironic that Cutter came to ballooning as if by a quirk of fate—as he explains in the interview.

Cutter grew up in an aviation family. Both his mother, Virginia, and father, Bill Cutter, were pioneers in the aviation business. His father started a fixed base operations (FBO) in Albuquerque, selling fuel, storage space, aircraft parts, maintenance, charter, flight training and aircraft sales; then he married the governor's daughter and taught her how to fly.

Sid served five years in the Air Force, flying fighter jets and the newer transports. He ran Cutter Flying Service after his father died in 1963 until 1974, when ballooning took hold of him and directed him as if by the wind itself. His brother Bill is now chairman of the board and Bill's son and daughter are now at the helm.

In 1978, Sid built and tested two thermal air ships for Anheuser Busch, which were twenty-nine thousand cu. feet, larger than Goodyear's blimp. No small accomplishment. Cutter's most endearing contribution to aviation comes from being the founder of the Albuquerque International Balloon Fiesta, the largest such event in the world, as well as from producing the first World Hot Air Balloon Championships in 1973 and again, the second, in 1975. From those humble first-year beginnings with only thirteen balloons participating, to where some eight-hundred-fifty-plus paint the Albuquerque skies with color, beauty and majesty, the event has become legendary, Today, Albuquerque is considered a world-wide mecca for balloons and balloonists. Still, this humorous and modest man is quick to share the credit. He says, "I'm pleased to get the festival going with such success, but I don't take credit for how much success. It's a bunch of hard-working folks who did it." The Fiesta is now run by a board of directors, and with the help of hundreds of volunteers, has become one of the country's most prominent festivals. Kodak claims more film is sold in those ten days than for any other event, including Christmas.

*How'd you get into ballooning?*

I started when I got a balloon for my mother on her birthday. I always gather things she could use like "glass bottom boats" and a happy Hote (standing Buddha), etc. The family had been in the aviation business for forty-two years at that time, so we thought we would throw a party for all our current and past customers. I wanted to decorate the hangar for the party. I had thought of using old WWI biplanes but they were very delicate and quite expensive. Then I stumbled on the idea of using balloons, since they were used in WWI, even though I'd never even seen one. I thought of renting one, but it seemed difficult, and because balloons cost less than five-thousand dollars, I figured I'd tie it up afterwards outside and use it as an advertisement—a gimmick which shows you how much I knew about balloons.

So I bought this balloon, but it wouldn't fit in the hangar. We finally managed to get it squashed in, and it formed the centerpiece of the party. The party was a success with five bars manned by French, British, and German costumed bartenders, an oompah band and a movie of *Those Magnificent Men in their Flying Machines* showing in one corner of the hangar. After the liquor flowed a bit, my brother Bill got to telling people we were going to fly the balloon in the morning. I heard this and said to Bill, "I don't have any idea of how to fly a balloon." Bill said, "Well, you've got a license, don't you?" (At the time all you had to do was show them your flight physical and they'd write you out a balloonist license.) I said, "Sure, I have a license"—and so did he and so did Mom—but that didn't mean anything. The next morning at six, my brother's beating on the door. I'm in my bathrobe but came to the door. He said, "Let's go ballooning." I said we couldn't because we didn't have any fuel. "Yes, we do," he said. "I stopped and got some fuel on the way home." "Oh, I said. "Okay. "I threw on my pants and we decided we'd raise the balloon on a tether (ropes) and learn how it works. Then we'd go to breakfast. Since it was her birthday present, I put my mother in and we flew it up on a rope, and I also learned how the burner worked. Then we took it all the way up to the end of the rope, and that was nice. I figured we'd give my brother a ride, then shut it down and go have breakfast, but there was a lack of communication and I got to the end of the rope with him and the ground crew let it go. Now we found ourselves free flying and drifting toward power lines: I found myself having to learn to fly quickly. I had been in the aviation business, so I understood some basic concepts like kinetic energy and delayed reaction, which occurred because of the large mass of the aircraft. The experience wouldn't have been all that bad, but I suffer from acrophobia. Here I was sitting on the tanks with my nose over the edge of the basket, pulling on the chain to get the burner to work. Finally, we got to three hundred feet, but I was scared to go up any further, and I didn't dare go any lower because of the power lines below. Fortunately, it was one of those perfect days for ballooning. Otherwise, I wouldn't still be doing it. I'm not a daredevil. But it was nice to be perceived as one!

Before long, I noticed I was drifting toward my house. We'd taken off from the International Airport in Albuquerque. At the time, the balloon was only drifting at about four miles per hour, but I didn't know how long the fuel was supposed to last, because I didn't intend to go flying. At three hundred feet, it looks like there are nothing but telephone poles ahead. Even though I'm a native and knew there are lots of open spaces, I was quite apprehensive. We were coming up on my house, so we used the VHF radio we had to call a satellite airport's UNICOM. I asked them to call my wife and tell her to look out the window to the south and see our balloon. The University Golf Course was located near my house, and we managed to move away from the power lines long enough to land. Some duffer said I made him get a nine on the hole.

When I picked up the balloon in Sioux Falls, I asked the factory man about balloon flying. I asked, "When you land a balloon, does it ever tip over?" The guy is from South Dakota, where the wind almost always blows, so he gets kind of glassy-eyed and says, "Always." This time we came in and landed and it didn't tip over. It just stood upright. I have to come up with some idea of how to get it
down. Five thousand people show up, along with five police cars. My lawyer (who was driving by) shows up and I put him in the basket so my brother can get out. My brother then pulled on the towrope to lean it over, and I pulled the "rip line," which makes it deflate, and it did.

In those days, you needed tools to take a balloon apart, but of course, the ground crew had the tools, and they were nowhere to be found. It turned out they had gone to breakfast up on the hill about three or four miles away. They watched out the window as the balloon drifted along. Pretty soon they quickly paid their bill and sped to where they thought it was. Unfortunately, we were in a draw, so they couldn't see where we landed, which turned out to be only a block from my house. I footed it over to the house, but nobody was home. I called my mother at the airport. I said, "There aren't any of my crew here." She said, "They're running around looking for you, so I'll tell them where you are when they call back." Currently, I'm at my house, but nobody is here either." She said, "Your wife is here at the airport." I said, "Really, let me talk to her." When she answered the phone, I said, "What are you doing there?" My wife said, "I got so excited seeing the balloon that I wanted to see you land." I said, "Well, I'm a block and a half from the house." She said, "Well, you always go to the airport to land." Anyway, that's my first balloon flight. From there we proceeded to go out into the open spaces and teach ourselves to fly. That was in 1971.

*What was the hardest thing for you to learn in those days?*

In those days, balloons didn't have much power—not much burner output. Back then, you had to anticipate the balloon a lot more than you do now. Then you had about four million BTU burners, but now they put out 22 million BTU's so that's a lot of difference. The power in balloons is really from brakes to stop a descent. For example, in those days, sometimes you'd burn for one thousand feet trying to get the balloon to stop. That would make you a little apprehensive.

*Any other tight situations flying in those days?*

Balloons are so simple that they're hard to get in trouble with. You always fly them in good stable air, because you can't get them inflated in the wind. On the other hand, if the wind comes up while you're up there, you're going to have to land, as fast as it is blowing. You'd have exciting landings sometimes. In New Mexico where we balloon's envelope is about two hundred pounds, plus the gondola weighs about two hundred pounds. Then there are eighty or so pounds of tanks with eighty-six pounds of propane. Thank makes the overall fueled weight about 570 pounds.

*Give me a wild flight.*

I don't have another wild flight. I do have some that were very awe-inspiring. For example, I've flown out of the Grand Canyon. The Grand Canyon flight was a multi-media project for Anheuser-Busch. The theme of the presentation had to do with how all the different ethnicities have all the space in the world. In flying, it is easy to get down. Of course, we try to get down on the road, because you only want to carry the balloon out once. To us, "lighter than air" is a joke.

*How much does the average balloon weigh?*

The average size sport cultures in the U.S. We also did a flight out of a bayou in Louisiana. But this one we flew out of the Havasupai Indian Reservation. The scenario AB wanted was that the kids could see the balloon, but the adults couldn't because it was a kind of fantasy. When we were flying, we were down in the gondola, hidden from the helicopter cameras. Anyway, we flew up out of the canyon to about ten thousand feet and tried to find a wind that would blow us out of the canyon. But there wasn't a wind, so we had to land back in the canyon. We got rescued by a helicopter and had to go pick up the balloon the next day.

Another flight was across New York City. In fact, I could have spit on the Empire State Building. That was awe-inspiring too. On that flight, there was a wind going from Manhattan toward Long Island down on the surface, but at one-thousand feet, I was heading for Ireland with no place to land. This was one of those deals where you'd rather die than make an ass of yourself, rather than go down below one thousand feet, to go out to Long Island. I lucked out and ended up putting the balloon right there on the beltway next to Jamaica Bay. Luckily, the wind got slower lower, and I was able to land in a small space. But when I got down, a helicopter cop, who'd been chasing us across Manhattan, arrested me. It seems you can't land anywhere in NYC unless it's at an airport or it's an emergency. To me this was a normal landing, but the cop kept trying to get me to say it was an emergency. I didn't realize the implications, so I kept saying, "Hell no, it ain't no emergency." Anyway, I got a two-page spread in the NY Daily News, which showed the cop with his foot up on the gondola, writing me a ticket

I was in the Fourth World Hot Air Balloon Championships in Sweden in January 1979, and the temperature was twenty-seven degrees below zero. Further, there were only four hours of daylight per day. If I hadn't been sponsored, I wouldn't have stayed. In that competition, our little balloon company took first and second place. I took the second, because I got a big penalty for not unfurling the marker right. Balloonists have this marker that weighs less than one hundred grams. It has a six-foot tail on it and has to be completely unrolled when you throw it out. I pulled out about three feet of it and didn't unroll the last bit. The judges hit me with a penalty that cost a bunch of points. That knocked me out of first. My partner, Paul Woessner, won and became a two-time world champion.

In Europe, all the roads go to the center of the town—like spokes from a wheel. In the earliest days, they designed towns this way so they could be better protected. Anyway, if you're going to cross town, your chase crew has to go back to the center. If they pick the wrong spoke, they may never find you.

*You raised the question of ego—admitting you're wrong. Talk about that.*

That's the thing that gets a lot of pilots in trouble, because they don't want to look dumb. Ironically, by not looking dumb, they may kill themselves. It stems from wanting people to respect you. If you admit you've done wrong, you risk losing respect. You just hope people won't find out and keep pushing the envelope until you end up having a real problem. One time I flew from the Bahamas to Florida, which is about sixty miles over the ocean. It was part of a balloon race, "The Bahamas to

Florida Balloon Regatta," which I put on for three years. Anyway, the first time we did it, I had to promise that the balloonists wouldn't fly into the Miami controlled airspace. We were actually flying from Bimini to Fort Lauderdale, but it turned out the wind was heading right for Miami. We had told the authorities that we would ditch, if necessary, but when I told the balloonists to, they wouldn't do it. As a result, they flew right down the final approach course of Miami International Airport. The air controllers there got a good lesson on how to control airplanes around balloons since balloons don't have rudders.

On this flight, my brother crossed one of the boats down there. Being a wiseguy, he asked the boater, "Which way to Ireland?" The boater pointed the opposite way.

*How many balloons did you have in the race?*

I think I had fourteen the first year. The race was real successful. We ran it three times. Then a local guy down there took it over. Later, I went back as a participant. We were flying and the trade winds quit. The thing is the trade winds don't ever quit, but this time they did for something like seven days. When the winds continued, the Balloon Meister said, "Okay, let's go." I said, "Well, where are all the chase boats?" He said, "They got bored and went home, but I'll call them and they'll meet you halfway." Sure, they will, I thought. Anyway, we took off and got about halfway there and the wind quit again. We're halfway between the Bahamas and Florida. With no wind, we slowly ran out of gas. I have my wife, Jewel, flying with me, the other guy has his wife flying with him. His fuel was also low, so he had his wife bail out at 30 feet into the water. A fishing boat picked her up.

*Were they having marital problems?*

No, but maybe afterwards. (Laughs) I couldn't throw my wife out because I didn't have a boat below. I ditched, but after landing on the water, the wind came up. There were three of us who had to put down in the water. Meanwhile, I tried to attract some boats, but there weren't any to attract. One fisherman came by and my wife hollered, "Mayday." The guy just waves. The coast guard finally showed up and picked us up.

*What's the feeling of ballooning mean to you?*

It kind of gives you a superiority complex, I guess. After all, you see those little people down there on the ground, all wishing they could be up there with you. What they don't know is that you wish you were down there, because a wind just sprung up. As for a profit-making venture, I haven't made anything but beans out of ballooning, but it has been rewarding personally. I helped get the balloon fiesta started in Albuquerque, New Mexico. About 850 participate in it. It's more of a rally, but contestants have received $15, 0000 balloons, plus things like diamond rings as prizes, etc.

*Could you be a poet?*

You must mean in relation to the beauty of ballooning. I don't fly along and recite poetry or pick leaves off of trees. That's not my way of doing it, but I know a lot of people who do. I'm one of those guys who flies to a target or lands on a spot. I love competition.

*What did you have to unlearn from other types of flying?*

In other flying, you pull back on the stick and it goes up. That doesn't happen in ballooning. A balloon has its own mind, especially if you're in a thermal. Of course, you fly in stable air if you can. If you get in thermals, it turns out that God starts flying the thing for you. Of course, gas balloonists are about half nuts anyway. Though I'm a gas balloonist, I try not to be fanatical about it. Most gas balloonists take off in the most beautiful weather, then end up flying in thunderstorms. They'll have landings where the wind is doing 30, 40, or 50 m.p.h. That hurts. I know because I've been there. You try not to hit the ground too hard, especially on a frozen field. You try to slide it. As long as the balloon is inflated, it acts like a big spinnaker, and it keeps going until it deflates or you fall out.

I was in a national championship once and it turned out there was what's called a low-level jet up about 300 feet above ground. The wind there was doing about 40 mph, but close to the ground, it was doing about ten miles per hour. That makes for a wild shear. I was flying a little balloon (small mass), a kind of racer. You know, the smaller the balloon, the more agile, and the less you can carry. Anyway, I was letting down the target to drop my marker on it when I hit the wind shear. Of course, what makes an airplane fly is the lift over the top of the wing, but if you get a strong wind below the balloon, it tends to pull the balloon down. This time it pulled me into a tree, and a tree limb hooked on my rip line and pulled the top of the balloon out. I'm sailing at about one hundred feet with no top in the balloon. The next thing bound to happen is that you're going to come down and so I did. As I came down, I went crashing into a big cornfield in front of me, and the envelope of the balloon lay over the corn. It got real dark in there amid these eight or ten-foot stalks. This was in Iowa, and they've got some big spiders there. Some are nine inches across. I don't like spiders, and I'm in this ten-foot corn with them. I've had a hard landing, but I'm fine. I decide I'd better get out of there and show everybody I'm all right, because the incident occurred right by my target in this competition, and my crew had seen me crash, so I came out and there's this guy there to take my picture. Of course, I look a mess. I say to him, "Dr. Livingstone, I presume?"

Every crash is embarrassing, and only a balloonist would understand why you crashed. Eventually, the farmer came out and said, "Don't worry about my corn. You didn't hurt enough to make any difference." It's pretty hard to be mad at balloons, because they're so pretty —hard unless you're a cranky old fart that doesn't like anything he can't eat or spend.

*What's the most unusual balloon you've flown?*

Probably that would be a balloon called "Sunstat," which is a solar balloon designed by Iranian, Fredrick Eshoo. It had two small electric model airplane motors with propellers in an attached tube, mounted at the balloon's equator. If you put some sort of air out of one side or the other, then it would turn. The balloon is a clear vinyl with nylon scrim on one side and black on the other. You could catch radiant heat, through the clear side when pointed to the sun, and it would go up. Turn the balloon around and put the black side to the sun and it would cool off—it's amazing. Furthermore, the colors of the ground you're going over will affect the lift. We had a burner on board but were not using it. We were doing a total solo flight, going up and down. I actually made a full, easy, egg-crate landing, just using the sun's power.

*What were the worst problems you faced in the early days of ballooning?*

Probably maturity.

*What was the worst thermal you've flown in?*

Thermals get pretty vicious, but as long as you don't run into an obstruction, you're not going to get hurt. Maybe if you had a real old balloon and hadn't kept it up, maybe it would tear and collapse. When I was learning to fly, we didn't know you should only fly balloons in stable air. We were just a bunch of airplane and helicopter drivers. Also, we didn't know that balloons flew better in the morning. It never occurred to us that a balloon could get uncontrollable until we flew through a few thermals.

I remember one time I stopped to change passengers, and a new passenger got in. As soon as he did, a thermal captured us. I never touched a burner from Albuquerque, which is at 5280 feet to when it got heavy enough for us to come down at 10,000 feet. That shows the power of the thermal. It could pick up all that weight, even without the hot air added.

One of the wildest times occurred when I flew out of Busch Stadium. It was at one o'clock in the afternoon, and was one of those times you'd rather die than make an ass out of yourself. We had a brand-new Budweiser balloon to inflate just before the St. Louis Cardinals baseball game. We were supposed to deflate it after inflation, but the players came out onto the field too soon. We were still on the field and didn't have time to deflate it and pack up, so I went ahead and flew up. It was now one thirty and thunderclouds occupied all quadrants. There was also a fifteen mile-per-hour wind over the ball field. In flying out, I figured I could just pop out of the top of the stadium. Then I'd go over and land by the St. Louis Arch. That was in the days when balloons weren't very powerful.

I didn't want to end up in the stadium crowd, so I needed to get through that wind shear. That meant I had to hit the top of the circular stadium going pretty fast. Certainly that's not the way you'd do it with the power now available. Anyway, I popped up too high to go back down to the Arch, and by the time I got the balloon to come down, I'm floating in there around all the skyscrapers downtown. That wasn't a good idea, so I flew back up above the height of the buildings but started heading toward the airport. Next, I got into some wild thermals, and they were taking me to the base of some thunderstorms. I'd swear I was going to fly into the clouds. But I did manage to come down some before then. Still, there was no place to land. I went by the same cemetery three times, and the third time I said to myself, "I'm going to land, even if it's sacrilegious." But I just missed the corner of the cemetery by a little bit and couldn't land. Soon I'm at the middle marker on the ILS (Instrument Landing System) to Lambert Field. At that moment, a TWA passenger plane came by and has to go around me in order to line with the runways. The tower said, "What are your intentions?" I said, "As soon as I can, I'm going to land this SOB."

I still have to find a place to land. I came up on another cemetery that's just off the airport. As I descend, I come to a big spruce tree and grab a hold of it till I can't hold on any longer. Then I let go and I pull the top out and land in the cemetery. As soon as I get the balloon packed up and into the bag (I haven't yet gotten it in the truck), hail begins to come down big as baseballs. I jumped in the truck to keep from getting killed. I could have easily died.

*Any forced landings?*

I had a double engine failure in a snowstorm at night once. I was flying a Twin-Bonanza, which is a pretty big airplane, coming back from Phoenix. It was absolutely smooth air in the clouds. The plane started losing power. I expected something like carburetor ice, but this aircraft has carburetors, which don't get ice in them according to the factory. It turns out the problem was that I was flying in super-cooled air and I was getting impact ice. Anyway, I ended up losing both engines, so I broke out two thousand feet below M.E.A. (minimum, en route altitude), which means I should have hit the ground. When I broke out of the clouds, it was night and terribly dark. There was no horizon. I looked out and saw Rte. 66 below. They were just building Interstate Highway 40. I was looking at Rtes.66/40. It was Sunday evening, so I thought I'd try to set the plane down between the traffic. There was no doubt I was coming down. I had both engines feathered. I kept seeing these car lights going off and coming on. I figured there must be an overpass there. As I'm about to make my final commitment, I see this truck turn off over the overpass. I figure that's the place to land.

I crank over into a sixty-degree bank, drop the landing gear and the flaps. I actually roll out, touch down and stall at the same time. I've landed on a sort of built-up dike road. As I put on the brakes, I see this sign on the left that looks as if it won't clear the plane. Because I can't get off, I move as far to the side as I can get, and miss the sign by about an inch. Then I look up and there is a bridge that this airplane won't fit across. I manage to get the airplane to stop. I'm about 50 feet from the bridge. To tell the truth, I had to tinkle like you can't believe. My first wife and mother are with me, so I crawl out over them to get outside. I just get started when Mom says, "Sid, here comes a truck." So I scramble back up the steps and flip on the master switch and the rotating beacon starts going. The guy in the truck hits his air brakes and gets it stopped. About that time, a pickup truck drives up. He came off Rte. 66 and asks if he can help, so we put a chain around the tail. I climb in to steer it back to a ranch road I'd seen. We let it roll down that road. Then a state policeman shows up and offers us a ride to Albuquerque. The Air Route Traffic Control Center phones my brother and says, "I guess it ain't you. We just had a pilot named Cutter go off radar. My brother says, "Dammit, he left me with all those debts."

Back at the airport, I saw Mom in her office and she's glassy-eyed, drinking Jack Daniels out of a quart ice cream cup. I said, "Mom, are you all right?" She says, "Come in and have a drink with me." She went on to explain that in the old days she used to deliver Eagle-Rocks and Stinsons (old airplane types) from the East Coast to Albuquerque, and that what had just happened was her first forced landing. This came as a surprise to me, because I'd had about fourteen at the time!

*What about your dad?*

Dad's license number was 1239. He figured the new CAA (Civil Aeronautics Agency) people would never get licensing of pilots through, so he didn't go get his until later. But Dad was quite an aviator and barnstormer. In the beginning, he probably flew maybe two hours before he was on his own. One time he told me he joined a flying circus. His job was to haul gas around in the front seat for the other planes. So they flew formation into some town, and all the other pilots start doing a loop. My dad had never done a loop, so when he got upside down, the gas cans started to shuffle out of the front seat of his open cockpit, so he jerks back on the stick and the cans slide back to the seat. Sometimes he said they'd come into a town and throw out these bombs to make a big noise, to attract people for rides, but they accidentally threw a bomb down the sheriff's chimney.

# Of Props and the Hummingbird

## Ray Hegy

His father was a blacksmith by trade, his grandfather an expert at the steam engine. Ray Hegy didn't go into either, but he did find himself making early-day wooden propellers for Hamilton Standard and engines of a different sort in the field of aviation.

Hegy didn't begin taking instruction in flying until 1927. He was working for Hamilton Aero Manufacturing Company, which later became Hamilton Standard. He had started at Marquette University, but dropped out when he realized he was "woefully" deficient in mathematics." At about this point, he saw an ad in the *Milwaukee Journal* advertising for expert cabinetmakers to work on airplane propellers. He applied, and on the basis of having made propellers for ice boats, got the job. Sometime during his three-year stay there, he insisted the company leave the dark stain off of propellers and go instead for a natural wood look, a tradition that continues today.

Ray Hegy has held a variety of jobs in different facets of the flying business (including seventeen years of aerial surveying), and one is hard pressed to describe his as anything but a colorful life, the stuff of serial adventure.

Hegy built his own plane, which a friend suggested he name El Chuparosa, or "Hummingbird." The suggestion seemed to be validated when a hummingbird actually flew in the shop and fluttered about in an effort to get outside. Unable to do so, it settled down on the plane. At the time of interview, Hegy no longer flies, and his plane is on display in the home-built section of the Experimental Aircraft Association's Air Adventure Museum in Oshkosh, Wisconsin. Recently Ray was written up in *Sport Aviation* and honored in his hometown of Marfa, Texas.

Ray Hegy was born in 1904 in Milwaukee, Wisconsin, and when I interviewed him at age ninety-one, I found him still vital and active. His company continues, too. Ray started Hegy Propellers in 1962, which later was run by his grandson, Jim Corder. The company makes new propellers and restores antique ones.

*When did you first get interested in airplanes?*

I saw my first plane in 1912, when I was eight years old, at Saskatoon, Canada, when I saw the Curtiss Exhibition Company. It was in much later years, in 1937, that I found out who flew the airplane. It was

Bob Saint Henry, an actor, whose real name was W.C. Shaeffer. He was from Sydney, Nebraska, where he retired. I read in the *Sportsman Pilot* about this guy who "flew with great success in Saskatoon in 1912."

*What did you think when you saw the show?*

It must have been early in the afternoon because I remember the sun glinting off some part of the airplane. I later found out it was the engine. It had a Roberts 50-horse, four-cylinder, two-cycle engine. The cylinders were cast aluminum and highly polished. (They were really made for marine use.) Anyway, that night after coming back from the fair, my mother gave me an empty Kellogg's corn flakes box. I made a model of the airplane from it. Mother had made some special paste of flour and water for gluing the parts of the plane together. By cooking it, she made a much better adhesive. (This was the same stuff we used to hang wallpaper.) So I made this model after she cooked a little of this in a tin can cover. I even had a guy sitting out in the front with the tricycle landing gear and featured the double triangular-shaped elevators and stabilizers. It was quite a deal. That's where I got my first interest in aviation, and that's never been diluted since.

*You were involved early in making propellers.*

After you've made a few hundred, you can do it with your eyes shut. But you had to have the equipment. You had to have an iron surface plate to mount the propeller. You'd use a protractor, one that's graduated to 1/10 of a degree. A navy inspector, who inspected Hamilton's propellers, gave the one I have to me. It was used in World WarI, I think.

About 1927, we started making metal propellers. I helped make Hamilton's first metal propeller, chopping it out of solid billets. Later on, the Aluminum Company of America made forging dies. Then the blades came out forged to approximate shape. All you had to do was to turn down the shank that went into the hub and grind or modify the rest of the surface.

*How long did you do that?*

I did that for about one and one-half years. My lungs got so full of that metal dust that I lost all appetite. I couldn't eat. I subsisted on braunschweiger and rye bread. At the time, I boarded with a family on the other side of town, and the woman was a very good cook. Anyway, she got so tired of making those sandwiches; she couldn't stand the sight of them anymore. I'd get a pint of chocolate milk to wash it down, but that's all I could eat. I'd come home each night, hungry as the proverbial bear, and once I picked up my knife and fork, I couldn't eat, so I had to quit the job.

*What happened then?*

A man out in Hartford, Wisconsin, where I used to live, said he'd give me $500 if I'd teach him to fly. Well, I'd been taking instruction at the county airport in Milwaukee. There he said he'd buy an airplane, if I'd teach him to fly. So he did. He bought an OX Waco 10. I flew that thing for ten years till 1938, when a company from Chicago came up to where I was in northern Wisconsin. They had a Bellanca Skyrocket with a 450-horsepower Pratt-Whitney Wasp engine in it.

*They offered you a job?*

Yes, they had had some trouble with their pilot. He was an air show celebrity, or thought he was. Every time there was an air show, he lit out. So these people had gotten tired of that and asked me if I could fly the Bellanca. Oh, boy, did I want to! So I flew that plane for a number of years until the war came along. My job was aerial photograph, exclusively.

My boss had been commissioned in the air corps reserve from World War I. He flew when they still used glass plates to take aerial photographs. He learned his trade in aerial mapping in Europe in World War I, and later he established his own aerial mapping company. I worked for him until he had to report for duty in World War II. That left me unemployed.

*What next?*

About that time, I was offered a direct commission by the U.S. Air Force, which used to be the air corps. They put me in a mapping squadron at Bolling Field. Next, they sent me down to Mexico to carry surveyors down there for the purpose of making new maps. I was essentially the chauffeur and didn't do any of the observation work. The group carried instruments for observing the stars at night from ground positions. They used these to compare with photos taken from high-flying bombers. After I finished up in Mexico, they sent down to Brazil. I was down in the Amazon for two years, flying a Grumman Goose.

*Any wild adventures in Mexico?*

I had a million dollars' worth, but I wouldn't give a nickel for another boatload! In Mexico, we used rudimentary airstrips that had been established by the Wrigley Company for hauling chicle, which was used as a basis for chewing gum in those days. In Mexico it went fairly smoothly. In fact, I was only down there for two and one-half months. I used those rudimentary fields, and some were on the side of a mountain. You landed uphill, and you set the brakes so you wouldn't roll back. Then you'd turn around and take off flying downhill, regardless of what way the wind was. I flew in a 1928 Fairchild with a beautiful Wasp engine in it and was the envy of all the Mexican pilots.

*Any forced landings down there?*

None, but I did later in Brazil. In Mexico, I had one of the finest compliments paid me I've ever had. It happened to be the time of year when they were burning off the jungle for planting new sugar cane and the alike. There was a lot of smoke. At the time, we were flying out of Tuxtla Guttierez in the state of Chiapas. A Mexican pilot said, "You won't ever find that place. I'll have to show you where it is. You'll have to fly with me, and I will show you all these places." He had a 1928 four-place Travelaire monoplane with a 240 horsepower J6 on it. (I think they only made four of them.) So he took off and, after a while, he asked me if I'd like to fly it. I was sitting in the front seat with him—we had a load of freight up to the roof. I said, "Sure." I flew it around awhile. After we got back that night, we had a couple of beers, and he was talking to the pilots and mechanics around there. He said, "This is Captain Hegy. I had him fly the airplane, and as soon as he took over the controls, I knew I had a pilot."

Later on, down in the Amazon Valley, I lost an engine. It was my first solo water trip. I hadn't flown a Grumman Goose before. I was put in that plane at Laredo. At first, I couldn't even taxi it. The Goose has a narrow gear, and the tail is high. In a crosswind, it's almost unmanageable on the ground. I finally got the hang of it by the time we got to Rio. By then I was able to fly it alone.

My first assignment took place around the Sao Francisco River, which empties into the Atlantic near Rio. Anyway, we followed this river up to its headwaters, and something happened with an engine. It may have broken a piston. Here I was on my second solo flight in the airplane, and it was getting dark. We were maybe thirty to forty miles from Jozeiro, where we were based. Well, I didn't make it. Though the airplane was supposed to fly on one engine at 6000 feet, I found I couldn't hold it. The propeller on the malfunctioning engine had stopped crosswise too. There was too much drag. I tried, but when I used all the power on the good engine, I couldn't hold it straight.

I got down below the riverbank. The banks stood maybe twenty feet high. So I ended up skimming over the water. I knew I couldn't get up over the embankments. The moon was up, so there was good light. My copilot, Captain DuBois, from Dubois, Pennsylvania, said to me as we flew along, "Golly, you pretty near hit that canoe-load of people." I said, "That was no canoe-load. That was a rock ledge." We had just skimmed over it and plunked down on the other side. When we did, the plane came to rest on a rocky bottom. As it did, it sprung a little leak in the keel, right forward of the step. My mechanic hit it with a ball peen hammer, and it straightened out. It didn't leak again all the time we were in South America. There was a boatyard there serviced the sternwheelers that navigated the Sao Francisco River in high water season. We were right there, and they pulled us up out of the water, with the nose out of the water and the tail on the bank. We notified our base at Belem, out on the East Coast, and he sent out a twin Cessna C-78 with a couple of mechanics and tools to install a new engine that they planned to send out on a C-47. The next day they sent the engine, landing on the airfield. The problem was we had to get the engine down to the river. We used a sailboat that had a ragged, old sail on it. When we tried it, the boat sank down to where it had only about four inches of freeboard. I was afraid we'd sink the boat, but we got over to the other side. The boatyard rigged up a wooden timber scaffold with not a bolt, wire, or nail in it, tied together with rope. They had a chain hoist, and we got the old engine off, slung it on the boat, took it across the river, and loaded it on the C-47. Later I found out they couldn't even pull the cylinders off the old engine. Some of the cylinders were actually bell-shaped from all the stuff flying around in there.

*How much did you fly that plane?*

Seven hundred hours. We got so we could fly rivers and streams, only three wingspans wide. We'd get down in there. The problem was the 100-foot trees on either side of the river. It was like landing in a canyon. Once we got a little dissuaded, when we landed in a place about 200 miles north of Manaos. At first, it looked like a nice straight stretch of river, but we didn't realize there was a stretch ahead with a bend in it. I knew immediately that I couldn't land there, so we elected to land on this straight stretch farther out. When we got down, we saw that there was a great big cedar tree leaning out across the river at about a forty-five-degree angle. I knew I couldn't get over it on the takeoff, and I certainly couldn't get under it. As you know, with a flying boat with floats out at the wing tips, you can't bank to make a turn on the water. Well, we were stuck. We had to make this turn. So we kind of skidded around this curve in the river, and all at once, the entire windscreen was covered with solid green. I knew we were in the trees. I said, "Oh, God." I never prayed better or more fervently in my life. I guess. Anyway, suddenly everything was clear, but in the float struts, there was a big piece of tree branch hanging. It was about 100 miles to the first habitation to land to check the damage. Then we took it to our base at Belem. It took them a week to iron out the dent in the wing.

*Did you meet any wild natives?*

Yes, we lost one survey party there up in northern Brazil. Because it was too narrow and crooked to get in with the plane, we had to send a party up by boat. They rented a batteau from a place called Moura up on the Rio Negro, above Manaos, to carry enough supplies for a week or two. So these four or five guys proceeded up this narrow river. Then they came to these rapids that they couldn't get though in a boat, so they transferred what they needed to the canoe. They only had a short distance to go and were making this transfer, when these natives appeared on the short. There we eight of them. The lieutenant in charge of the survey party got out his camera to take a picture, and the air was full of arrows almost immediately. They killed all the members of the party, except the boatman. (If you had a boat with power, you had to hire a boatman.) The boatman was a Negro and a very strong swimmer. He dove into the water and got away. Thirteen days later, the owner of the boat they'd rented discovered him, almost dead from starvation and exposure. He brought the story back.

*Did you have to go to the site of the killing?*

Yes, we had to go back and try to recover the observation. We found the remains of the lieutenant out on a rock in the middle of the stream and identified him by his sneakers and blue swimming trunks. The cranium was with it, and that was later used to confirm our identification through dental work. The piranhas had undoubtedly gotten all the rest of the party. We didn't recover the records, and I don't think the observation had been made. Anyway, it turned out the lieutenant was more interested in getting artifacts than doing the observations. In fact, he had a large collection of stuff at the place we stayed in Manaos. I have some of those things in my shop here—bows and arrows and a twenty-two-foot snakeskin, an anaconda.

When we found the remains of the lieutenant, we put them in an instrument box and carried it down to Manaos. He was buried there.

*That was pretty tough duty—tough emotionally.*

We had a military funeral, and the Brazilians were very cooperative, taking care of everything. We mounted guard with our military personnel in the funeral parlor. There was no embalming, and the odor was quite overpowering. We smoked to cut down the odor.

Observation points were nearly seventy to one hundred miles apart. Sometimes the place designated would be under forty to fifty feet of water because the photo was made during the dry season.

Later those same Indians were still up there. Our people rented a boat again, one with an old Swedish Archimedes engine on it. The engine was all worn out, so it ran in fits and starts. We flew surveillance over the party in a PBY and saw these natives. There were two canoe loads. One had five in it; the other had three—for a total of eight, as before. When we flew over, the Indians would duck under the trees. The trees came down to the water's edge. Anyway, when they were under this canopy of green, you couldn't see them. (We could have done them great harm, because we had firearms, including carbines, but we were not allowed to use them.) On the way back the engine pooped out, and the Indians were coming. Captain DuBois was trying to start the engine. Just about the time it started, the Indians caught up with them, but they got away. We later picked them up in the PBY at the place the equipment had been rented. We found out later that these people were not native Brazilians, but came in Brazil during the wet season from Columbia. One day we came into a clearing and flew over their village. It had a large, circular, thatched house, beautifully made.

After two and one-half years, I came back and went to work for my old employer. During World War II, he had been instrumental in installing air bases for the ferry command. He was in a PBY, taking off from the Saint Lawrence River, when they ran into trouble. The water was too rough, and the airplane broke apart. John almost lost his life right there. I guess it did permanent damage, because his heart failed him walking into a hotel lobby in Los Angeles a couple of years later. When that happened, I was out of employment again.

I took employment with an aerial survey company in Midland, Texas. These people had decided to obtain a plane for high-altitude photography, taking pictures at twenty to thirty thousand feet, which was quite new in those days. Anyway, someone sold them on the idea of getting a P-38. I'd never flown a P-38, but here was this beautiful, brand-new plane up in Kingman, Arizona. So these folks sent me up there to fly it back. The P-38 we were going to get was an F5 model. The plane was complete except for cameras and radar altimeter. The cost: $1,250 and it had only fifteen hours on it. This was in 1946.

So I took an airline out of El Paso to go get it. I met a man on the plane and he said, "Aw, you won't have any trouble. That's an old man's airplane." I said, "I don't want it making an old man of me." He replied, "You won't have any trouble." And I didn't. But four guys flew up to Kingman in a Cessna Airmaster to get two of these P-368s. But when their pilot went out to get this airplane, he couldn't even taxi it. He didn't have any idea at all. The crew chief had given me about thirty minutes of cockpit time, showed me where all the levers and switches were, so I was all right. He had me fire it up. These planes had been out there for months, so the wheels sank down in the sand. The crew chief tried to horse this think out with the engines, and he ruined the prop with all the flying gravel. I said, "I don't think I want this airplane." He said, "Which one do you want?" I picked another after looking over the logbooks. They pulled it out with a tractor and checked me out at headquarters and refueled me. I remember it took 4440 gallons of gas. It turned out there was no clock in it. I noticed that and he said, "There isn't a clock on the field. But if you've got fifteen bucks, maybe I could find one." So I gave him the fifteen, and they filled up my tanks with gasoline. They also filled up the nose section (where the cameras would be) with spark plugs and all kinds of stuff and never charged me a dime. But these other guys wouldn't buy a clock. They had to buy their gas.

I heard later that these people got the P-38s to Tulsa. No place for the cameraman, except in the nose. It's a tricycle-gear airplane, damned if they didn't stick the strut up into a cameraman one day, and kill him.

*Did your plane work for you?*

We tried to fly ours, single-place, but to operate a photographic airplane, you have to have one person to fly the airplane and another to run the camera. We rigged it up as a single-place machine with levers to control the camera, but we couldn't make any acceptable photographs. I flew it about twenty hours and quit. I came home and started a furniture repair business.

*How many years in aerial survey?*

From 1938 to 1955.

*What's the toughest thing about aerial survey work?*

For one thing, you have to know how to fly straight. Very few pilots can fly straight. I mean straight. Now I learned to fly straight, but I wasn't always in the right place. I've had more reflights than anyone in the business, I think. It was kind of discouraging. On the other hand, they always sent me out to the roughest part of the country—to the mountains out west, someplace like that. I could fly straight all right. For example, I flew the entire city of Houston without a re-flight. They didn't have to redo a single picture.

*What year was that?*

1954, I think.

*Who was the biggest character you've known?*

I knew Eddie Heath, the foremost light plane pioneer. He acquired a lot of World War I war surplus and sold airplane parts. One was the Heath Parasol. I built one, too. After I'd flown his, I realized I didn't like it. The wing was so far above the fuselage that the fuselage wanted to go around the wing when you took off. Heath was a little guy. I weighed 125 pounds at the time, and he weighed 110. He was slightly shorter than I. I was packing parachutes for Curtiss at Glenview, Illinois, and used to fly over to Ed's field in Joliet, and fraternize. When I made my own Heath, I lowered the wing six inches, just enough to clear my head. To do that, I had to engineer a door in the side. I copied the truss from the structure of the Waco.

*Did he like your plane?*

He never saw it. I started building it in 1928, and he was killed, I think, in 1929. It took me two years to build the Heath Parasol. We flew the plane for 600-700 hours, and the guy who helped me with the engine soloed in it. I taught him in the Waco.

The man who taught me to fly always remains in my heart. He was Cassius B. "Cash" Chamberlain. He later went on the Northwest Airlines. About 1937 he said, "Come up to Minneapolis and get a job with Northwest." I didn't have any night-flying or instrument experience, but I went up. The Northwest people spent almost a day with me, which led "Cash" to say. "They sure gave you a lot of time." Well, nothing came of it, and I'm glad. A year later Cash was dead. A Lockheed Zephyr that he was flying caught fire in the air, and they couldn't get back to the airfield at Miles City, Montana.

# Beginning in 1929

## Fleming Cooper

Fleming Cooper was born in Columbus, Georgia, in 1909. When he decided to go to Georgia Tech, his uncle, who was his guardian, suggested he consider taking Textile Engineering since Columbus was a textile town. He entered that course and worked two summers in a cotton mill when he realized that those who moved to top management were men whose families were mill owners.

His first exposure to flying was when his Applied Mechanics instructor, who was an ex-Navy pilot, decided to buy a training glider and organize a glider club. Fifteen members of the instructor's class each paid $25, which paid for the glider. The fuselage was a four-inch-wide truss made of 2 X 4s with the wing attached to the top member, and the bottom member served as the landing skid, on top of which was the single pilot's seat. There was no seat belt so the pilot ran his pants belt around the back of the vertical member to hold him in the seat.

No member of the club had received any flying lessons, but they read a Navy pamphlet on "How to Fly" and on windy weekend days, the members of the club took turns sitting in the glider on the ground, that way they learned the relationship between the control movement and the glider response. Since most of them had never been in an airplane before, it is a miracle that none of the 15 members of the club "didn't bust ass." There were no injuries and equally amazing is that the members could get away with operating this activity without any liability coverage. "That was in 1929, before the heyday of today's Trial Lawyers," Cooper remarks.

Cooper and the others became airborne, by being sling-shot. This was accomplished by a long rubber shock cord, the middle of which was attached to a curved hook on the nose of the glider. One person would hold on to the tail of the glider while two people would stretch the shock cord in the direction of take-off until the person could no longer hold the tail. The guy in the pilot seat would suddenly find himself about 100 feet above the ground, with no training about how to safely return to earth. Looking back on this foolishness scares him even now, Cooper points out. He says, "I had never before held the controls of an airplane in flight and the only visual reference point ahead of the pilot was his feet on the rudder bar. There was no air speed indicator. This was the beginning of my flying career and knowing what I do now after over 13,000 pilot hours and fifty years as an active pilot, I consider that we were extremely foolish and lucky, lucky."

Toward the end of his sophomore year at Georgia Tech, a friend who had gone to the Air Corps Flight training gave Cooper a flight in his airplane and he was hooked on flying. When the Guggenheim School of Aeronautics was started at Georgia Tech at the beginning of his junior year, he switched from Textile Engineering to Aeronautical Engineering. A year later, he applied to the Navy and to the Army Air Corps for flight training and was accepted by both, and chose the Air Corps. He entered the Army Air Corps in October 1932 as a Flying Cadet and was in the first class to begin their flight training at the new Randolph Field near San Antonio, Texas.

*Tell me a story about the early flying days at Randolph.*

Before getting into it, I must set the stage by first referring you to the PT-3 cockpit. The instruments are mounted on the outside of the cockpit with the tachometer on the left; the large instrument on the right is the non-sensitive altimeter; next is the oil pressure gauge then the oil temperature gauge; the compass was also on the outside of the cockpit. Note the absence of an air speed instrument. Directly ahead of the cockpit is a big fat leather pad to save the pilot's face in case of a crash. No shoulder straps had been thought of at that time. The aircraft had the Wright J-5 engine (the same model that Lindbergh flew across the Atlantic). It cruised at 81 MPH and stalled at 48.

Proper tuning by the instructor of the student's senses of sight, feel and sound were the essential elements that the instructor used to teach the student how to fly. The instructor sat in front with the student in the rear cockpit. Communication was one way through a rubber tube called a gosport. (I must be getting old that word is not in my Webster's dictionary). On the instructor's end of this communication system was a metal funnel into which he could speak his words of wisdom; student's end of the tube was a "Y" joint that ran tubes to the student's helmet over each ear. Some instructors who were getting slow responses from their students would hold the funnel of his speaking tube out into the slipstream and the blast of air into the student's ears would immediately get his attention. The student did not have a funnel into which he could speak so he had to save his questions until after landing. I never used an aircraft radio until I started flying for Boeing Air Transport (became United Airlines in 1934), seven months after I graduated from Kelly Field in October 1932.

The PT-3 was a neutrally stable, or better described as a slightly unstable aircraft. The controls were very stiff mechanically and aerodynamically and a lot of muscle was required to perform most maneuvers. One had to fly it all the time and since it did not have a trim tab, pressure on the control stick was necessary most of the time.

Because of the absence of an airspeed indicator, we were taught to keep the nose a certain altitude below the horizon for glide without power. Since the aircraft was directional-neutral stable, we were told to detect skidding by the unequal blast of wind on our face. We soon learned to notice skidding by the sway of our body above the hips.

All approaches for landings were without power. We had to set up a 180-degree traffic pattern 400 feet above the elevation of the field and required to completely cut off the power on the base leg. From then on, we maneuvered much altitude, the proper speed and headed in the right direction. At any time and frequently during our 45-minute instructional period, our instructor was apt to cut the engine power and you had to make a simulated forced landing approach the best available piece of real estate.

After completing my first instructional period on landings with my miseable attempt to make everything come out even, I asked my instructor how do you know when to turn off the base leg onto the approach leg.

His response was a classic, but it worked. He said:

> You look at the field.
> Cut off the power.
> Drop the nose to the gliding altitude.
> Listen to the sound the wind makes on the flying wires of the wings.
> If the sound is too low a pitch, you are going too slow.
> If the sound is too high, you are gliding too fast.
> Look back at the field.
> Check the altitude of the nose.
> Listen to the wind sound.
> And keep repeating the above until you can say, "Now is the Time." Then make your turn onto the final approach leg. Keep your nose gliding altitude and the proper wind sound, until it is time to pull the nose up for a three-point landing.

The PT-3 had rubber shock cords on the landing gear, without any hydraulic shock absorber to reduce the after bounce, so every landing had to be a three pointer at stalling speed. Otherwise, the airplane would bounce like a rubber ball thrown against the ground. If you landed above the stall speed and bounced, the nose would pitch up increasing lift on the wings, which increased the height of the bounce. If you didn't quickly add power, you would stall too high above the ground and really make a hard landing. In spite of its faults, the PT-3 was an excellent trainer because it was difficult to fly.

*What is the worst advice you ever got from an instructor?*

It wasn't a bad comment, but kind of funny, I thought. The instructor picked out a field and said, "I want you to land on that mowed strip." It had recently been mowed and there was a haystack on one side. He said, "I want you to land on that strip and stop even with that haystack." I landed and, of course, those airplanes didn't have brakes in those days. You'd just pull the throttle back and come to a rest. Anyway, the airplane rolled about 20 feet past that haystack, which I thought was a pretty good performance with all the variables I had to contend with. But, the instructor didn't say a damn thing about it, anything like "good landing" or "good approach." He said, "Were you an engineer before you came into the Air Corps?" I said, "I went to Georgia Tech." He said, "No," and he said, "Then quit hanging sideways out of the cab."

*You mentioned before we started something about one of your fellow cadets missing the Rio Grande River. Could you relate that story?*

Before graduation, we had what was called a graduation cross-country. We went from San Antonio to Roswell, New Mexico, up to Ft. Lawton, Oklahoma, down to Roswell, then down to El Paso.

One cadet was really a super egotist in my opinion. He was always right and everyone else was always wrong. Anyway, when he got to the Rio Grande, he didn't recognize it as the Rio Grande. He didn't think he was close enough for him to be there. It looks just like a branch. In fact, the Rio Grande is a very narrow stream in some places. So he crossed over the Rio Grande and kept flying south into Mexico, where he finally ran out of gas and landed. He wasn't heard from because they kept him in jail that night. They apparently didn't know what to do with him. He didn't call in, so he was missing. The next day we organized a search party to see if we could find out what happened. We thought maybe he had trouble between Roswell and El Paso. What happened finally was the Air Corps lost a plane and we got him back from the Mexicans two days later. I graduated as a Second Lieutenant Air Corps Reserve October 1932 from Kelly Field, Texas.

My orders assigned me to eight months active duty to the 20th. Pursuit Group at Barksdale Field, Shreveport, Louisiana, which had just been completed. Times were tough then and our $125/month plus 50% hazard pay came from CCC Funds. And we had to take one months' vacation without pay. I loved every minute of my duty at Barksdale. Uncle Sam assigned to me a P-12 Boeing Fighter aircraft with my name on it and furnished a skeet range with free ammunition so we could learn to aim ahead of a moving target. We flew a lot of formation, did aerial gunnery and other fun flying and we were paid what at that time was big pay of $125 a month. Shreveport opened their arms to us; the Country Club gave us memberships without any initiation fee. We met many very attractive Shreveport girls and many of us including me were fortunate to marry a Shreveport girl. The local young men did not like us because not many were making as much as we were at that time. We were flying open cockpit aircraft requiring us to wear helmet and goggles. Our faces would tan everywhere except under our goggles, which made us look non-standard. In envy or jealously (maybe both) our competing local men referred to us as "THE FISH-FACED FOOLS FROM THE FLYING FIELD."

Since my orders were for only eight months active duty, I did not want to stop flying, so I applied to and was accepted by Boeing Air Transport as a copilot, flying between Chicago and Cheyenne. Initially, I flew the Boeing 80 which was a three-engine, all metal, low-wing Biplane that carried 20 passengers at a cruising speed of 120 MPH; later we graduated to the Boeing 247 which was the first all metal, low wing airliner; it carried 14 passengers at 165 MPH. After the Airmail Cancellation in 1934, the Army Air Corps carried the mail for several months with high casualties. After the Airmail debacle, Boeing Air transport merged with several other airlines and became United Airlines.

The first flight I had west of Omaha was with a pilot named E. Hamilton Lee, who was a noted airline pilot. When we landed at North Platt at 4:00 AM, it was raining, very hard, with poor visibility. After we left North Platt, heading towards Cheyenne we started getting into heavier rain and, all of a sudden, I saw a beacon light on the right, about 30 degrees above where I thought the horizon should be. I looked at the artificial horizon instrument and it indicated we were flying level, but this beacon light was above us. I reached over and said, "Say, there is a light to our right above us." He said, "That's right. It's supposed to be there." That light was on top of a hilltop and we were flying to the left of it in the valley. The pilots who were the survivors knew when to fly to the left or to the right of each light beacon. Back in the old days, the old timers used to fly the Columbia River Gorge when the weather was bad, it was like flying up a tunnel with the river on the bottom and steep banks on each side with clouds above. The light beacons were on various sides of the Columbia. You had to know which side of the light beacon to fly on or you would fly into a hill.

The mid 1930s was a period of great advancement in the airline piloting. Many of the older pilots were former barnstormer pilots who were good "Stick and Rudder Seat of the Pants Pilots," but were very deficient in instrument flying capability. A few had very little formal education, had bought a WWI Jennies for $500 and had learned to fly. They were "Barn-Stormers" and, when the airlines started hiring pilots, some qualified. An Air Corps friend who went with another airline told me that one time he and his first pilot were weathered in and while waiting for it to clear, the First Pilot said to him: "The first time I seen an airplane climb up a tree that dove into it."

Co-pilots spent many nervous hours at night with some of these older pilots who would follow a highway at night in bad weather going from one light beacon to the next if he could see it. We would unsuccessfully try to see the ground on our side of the airplane, occasionally look at the altimeter, and say to ourselves; we can't fly this low in this area without hitting something. About 95% of the co=pilots were Air Corps trained and we learned what to do and what not to do by observing the First Pilots. It was a pleasure to fly as copilots with the younger first pilots.

Of course, some old-timers had been progressive and kept up with the State of the Art. One night I was with one of the old timers and we had a line of thunderstorms ahead of us. He tried to stay underneath the clouds but could not and pulled up into the very rough storm clouds. He was doing a lousy job of instrument flying. I thought he was having vertigo problems and I asked him if I could help! He gladly turned controls over to me. Shortly after that, United purchased some of the first Link Trainers and discovered that several of their long-term pilots were very weak on instrument flying. This resulted in United firing 35 First Pilots in one day. I was at that time a Reserve Captain flying out of Portland to Salt Lake and I immediately started flying regularly as a First Pilot. This lasted about three months when a Reserve Captain higher than I on the pilot's seniority list saw me fly into Salt Lake as a First Pilot. He was then flying as a copilot and rightly requested to be transferred to Portland to replace me. I was then transferred to Oakland to train new copilots to qualify them for their instrument rating. After six months of this, I was assigned to the San Francisco-Salt Lake route.

In those days, I would say that 98 percent of the younger airline pilots were Army trained. They were almost without exception very good, because they flew with these old timers and knew what was good and what was bad. They learned the good techniques and rejected the bad.

*How did the role of pilot and co-pilot change from the thirties on through to the time that you returned to the Army Air Corps?*

When I first worked for United, procedure was very lax. We didn't have a checklist. Of course, there wasn't that much to do. The Boeing 247 didn't have flaps, but you had your gas, propellers, and landing gear. That was the checklist you went through. They devised what was called "GUMP" list. "G" was for gas; "U" was for undercarriage; "M" was for mixture and "P" for propellers. As aircraft became more sophisticated, a need developed to formalize the coordination in the cockpit; the checklist became elaborate and today airlines are using simulators that are so good the trainee is given a pilot certification by the FAA before he has even flown that type of aircraft.

I loved flying with the airline, but I had an unexpected reaction after flying as a First Pilot, after four years sitting in the right hand seat as a copilot and dreaming about being a First Pilot. When it happened, I realized that I was still doing the same work except being paid more for it; however, I realized that I had gone as far as I could up that ladder. More important is that our first child was on the way and, eating whenever you got hungry and sleeping whenever you became sleepy, did not seem to fit into what I considered a normal family life. I had always enjoyed being in the Air Corps as a pilot, so in 1938 the Air Corps announced a competitive exam, for Lieutenant Regular Commission. That was the fourth time I had become a second Lieutenant, which takes a lot of explaining as follows:

First Time:2nd. Lt. Army Reserve from Senior ROTC Vacated when I became a Flying Cadet (Enlisted Status)
Second Time:2nd. Lt. Air Corps Reserve upon graduation from Air Corps Flying School While flying for United Airline I became a 1st. Lt. Air Corps Reserve through Correspondence Courses
Third Time: 2nd. Lt. Air Corps Reserve when called back on temporary Active Duty in 1935. Returned to United Airlines.
Fourth Time: 2nd. Lt. Regular Air Corps through competitive exam.

Going back to the Air Corps failed provide a more normal existence and more time at home. How wrong I was. WWII arrived and I was away.

*What was the hardest adjustment to flight training at Georgia Tech?*

I didn't begin flight training at Georgia Tech. I took the Navy ROTC Aviation course, which was a ground course. It gave me navigation rules for the road, but was strictly Navy-oriented. I didn't get into flying until I went to the Air Corps flying school at Randolph Field in 1931.

When we first got down there, the younger group was hazed pretty hard by the upper class. I was the first class at Randolph, and our upper class was comprised of cadets from Brooks Field in San Antonio and Morris Field in California. They competed to see who could be rougher on the lower class. We went through a six-week "Do-Do" class. A do-do bird can't fly because it doesn't have wings, so we were "do-dos" until we learned how to fly.

*What was the worst thing that happened to you, in terms of hazing?*

I think the worst thing that happened there involved this upper class fellow, who was a wonderful guy. One Sunday morning we were walking to church and he was with us. They used to call us "Do-Do's" all the time. Well, he said, "Mr. Cooper, do you think the upperclassmen from Morris Field are a bunch of heels?"

I didn't answer that, but the word got around that I said the Morris Field guys were a bunch of "heels," so then they started working on me pretty hard. Randolph—being brand new—had a lot of building clean up needed around the dormitories. For about four weeks straight I was assigned to picking up rocks and putting them in a pile. But it all worked out fine. Flying was so much fun that all that hazing wasn't so bad.

It was interesting to me that old pilots—barnstormers—who came on with the airlines—had never flown with co-pilots. Later on, airplanes were single-pilot like the Boeing 40. You had one pilot, a baggage compartment, and a place for the mail. Some had a place for two or three passengers. But because the old pilots never had the opportunity to fly with other pilots—they weren't able to pick up good pointers from them. So the old-timers, with a few exceptions, were non-progressive. They got in a rut. That is how they learned to fly and that is the way it was. I had some pretty rough times sitting on the right-hand side, watching these guys at nighttime, fly closely to the ground.

*Any memorable barnstormer that you flew with?*

Jack Knight. He was the first one who flew airmail at night. He was a real good guy. In 1934, the airlines started getting Link trainers for the first time. A Link trainer was a very unstable device. You got into it and it felt like you were getting on top of a pivot. If you were not flying the thing, 99 percent of the time you would drift off and fall to one side or the other. You had to really fly hard. Well, they were starting to get these old timers into these Link trainers, and I realized that some of them were pretty crude pilots. A Link trainer had a crab that would track the direction that you were flying. There was a graph and, if you got off to one side, you got a slight "A" and, if you got off more, you got a stronger "A." On the other side, you got an "N." When you were on course, you got a solid signal. Anyway, these guys didn't know how to fly that very well and they didn't have a lot of experience doing it. There was a big natural gas tank southeast of the airport in Chicago and this pilot was making a landing in a Link trainer. He missed the field and flew right into that tank. That woke up the industry management and they realized they had some pretty lousy pilots flying the airlines. I was transferred to Portland shortly after that and they laid off 35 First Pilots in one day because of the Link.

The Link trainer showed precisely what they were doing because you couldn't argue with a red line on a piece of paper, wandering off, showing you were lost, and then ending up into a gas tank or something else.

I had just been checked off as a reserve captain at that time. They laid two pilots off in Portland, so I got a chance to fly as a First Pilot out of Portland. I was lucky. That lasted about two months.

*How did you feel about these people being laid off? The right thing happened, but you must have felt a little sorrow for them?*

Not for some of them. I figured that if some of them kept flying, I might end up in the side of a mountain. Co-pilots used to get together away from home. We generally ate at the same place. We would talk about the bad pilots and cite situations where they had done something stupid.

One guy I was with took off from Lincoln, Nebraska, and we were heading for Cheyenne, Wyoming. He took off and started flying in the direction of Omaha, which was northeast. He kept on flying to the right, instead of to the left to get back on course. He flew for 35 minutes northeast when he should have been flying west. Finally, I convinced him he was wrong and he headed west. When we got to Cheyenne, he said, "Boy, we sure had a strong headwind."

*What about memorable instruction? Any stories?*

At the primary stage at Randolph, we had to practice simulated engine-out conditions almost every flight. You wouldn't always land, but you would go through the motions of setting up a pattern and coming in to make a proper approach with the plane. In those days, we didn't have a speed indicator. We had to fly by feel, sound, and altitude. It didn't take long to realize when you were going too slow or too fast. I remember I asked the instructor how you knew when to make the final turn in the landing. He said, "Well, you put the nose down, listen to the wires and pick up your speed properly, keep your nose at the right altitude, look at the field, listen for a while, look at the airplane, look back at the field. You keep sizing it up, then you come to the point and you say, "Now is the time to make the turn." Just about that simple.

*What is the worst advice you ever got from an instructor?*

It wasn't a bad comment, but kind of funny, I thought. The instructor picked out a field and said, "I want you to land on that mowed strip." It had clearly been mowed and there was a haystack on one side. He said, "I want you to land on that strip and stop even with that haystack." I landed and, of course, those airplanes didn't have brakes in those days. You'd just pull the throttle back and come to a rest. Anyway, the airplane rolled about 20 feet past that haystack, which I thought was a pretty good performance with all the variables I had to contend with. But, the instructor didn't say a damn thing about it, like no "good landing" or "good approach." He said, "Were you an engineer before you came into the Air Corps?" I said, "I went to Georgia Tech." He said, "No I don't mean that kind of engineer. I mean a railroad engineer." I said, "No," and he said, "Then quit hanging out of the Cub."

*Before we started, you said something about one of your fellow cadets missing the Rio Grande River. Could you tell that story?*

Before graduation, we had what was called a graduation cross-country. We went from San Antonio to Roswell, New Mexico up to Ft. Lawton, Oklahoma, down to Roswell, and then down to El Paso.

This one guy was really an egotist in my opinion. He was always right and everyone else was always wrong. Anyway, when he got to the Rio Grande, he didn't recognize it as the Rio Grande. He didn't think it was close enough on his map for him to be there. It looks just like a branch. In fact, the Rio Grande is a very narrow stream in some places. So he crossed over the Rio Grande and kept flying south into Mexico, where he finally ran out of gas and landed. He wasn't heard from since they kept him in jail that night. They apparently didn't know what to do with him. He didn't call in, so he was missing. The next day we organized a search party to see if we could find out what happened. We thought that maybe he had trouble between Roswell and El Paso. What happened finally was the Air Corps lost an airplane and we got him back from the Mexicans about two days later.

*How did the role of pilot and co-pilot change from the thirties on through to the time you retired?*

When I first worked for United, procedure was very lax. We didn't have a checklist. Of course, there wasn't that much to do. The Boeing 247 didn't have flaps, but you had your gas, propellers, and landing gear. You went through that checklist. They devised what was called a GUMP list. "G" was for gas; "U" was for undercarriage; and "P" was for propellers. That was a pretty good checklist in those days. The good pilots on United would share landings and takeoffs with you. The guys who were the weakest would never give you a landing—I guess because they were not sure enough of themselves. On landings, the co-pilots would get experience in handling the controls in landings and takeoffs. The pilots who were not very sure of themselves—the barnstormers, some of them, not all of them — would never give a co-pilot a landing. He nearly always landed.

One time I was with a guy who would never give a co-pilot a landing. He would take off and give it to the co-pilot to fly the trip, and then he would nod off. I think he was a heavy drinker, but I'm not sure. Anyway, we got to Cheyenne and I made the decision to land. I was not going to wake him up. The heck with him! Just before I pulled the throttle back to land, he woke up and took over. He was as mad as hell. He was the same pilot who took so long to get from Lincoln to Cheyenne.

*Did some of the early guys you knew like to dress the part, say, of barnstormers.*

Not entirely. I never saw this kind of thing, but one of the pilots out of Cheyenne used to wear cowboy boots and dress the part. He would carry a gun in a holster on his side. In those days, the pilot and the co-pilot were supposed to guard the mail so they issued each pilot a gun in case you had to land somewhere and someone tried to take over the airplane you were on. One time we landed on Grand Island and the weather was so bad we couldn't go any further, so we decided to spend the night. We were hauling a shipment of gold and one of us had to stay awake all night.

The relationship between the pilot and the co-pilot was pretty good in most cases. Most of them had respect for one another and the good pilots would share landings almost every time.

I think things really improved. Nowadays, the checklist is really complete and they go through it. Also, they have a recording on what goes on in the cockpit, which they can play back any time. They can do this after an accident, for example. In 1936, the airlines came out with a minimum altitude between checkpoints. They did this to stop old timers from hugging the ground in bad weather. One pilot, who was noted for contract flying, flew from Chicago to Toledo at a low altitude. He didn't know there was an altitude reporter on the plane. The authorities checked it out and found out he was flying low altitude after reporting high altitude.

*What was the first accident you witnessed?*

We had a two-target gunnery and a ground gunnery. With the ground gunnery, if you came in too fast, it degraded your score. So people would try to fly as slow as possible, so they would have more time to get around on the target. This guy was overdoing the slow flying and he just stalled and spun in. The next accident I witnessed was a formation collision. We were training for putting on an airshow. My roommate was flying this airplane in and another fellow somehow got too close to the lead airplane, and somehow chopped a finger off at 600' above the ground. My roommate jumped out—he was in the lead airplane—and this guy didn't make it. A buddy of mine said, "That's the way it is. Here today and can't hear tomorrow." I wondered what did I get myself into.

Another time, I had a real good friend at United— one of the old timers—who flew into the side of a mountain, east of Salt Lake City. Forty people were killed. It affected me, but you get over it.

*Describe early formation flying, tactics.*

When I went to flying school, I guess it was good training but kind of hazardous, too. We had a three-ship formation. You could make gradual turns and stay in place. Of course, when you made a turn, this ship had to slow up a bit and that one speed up to stay in position. There was a procedure turn in training where the lead instructor would give a signal (gesturing), a 90 degree turn. When that happened, this plane knew he had to cross over on top, and he would come down. On the left angle, someone would go under and up. During the time you were crossing, you would actually lose sight of this plane and that plane. They had several mid-air collisions doing this. Then they went to an echelon tactical flying squadron leader. There was a radical change in what they used to do, because always the wing man would fly high. Sometimes in abrupt weather—and maneuvers when you crossed over—you would lose sight of the guy ahead of you, which is hazardous. Nowadays, the wing man flies low and the other flier can move around at will and still maintain position. It is much more flexible, a much more maneuverable situation today.

*What makes you a great pilot?*

I had an awful lot of good training at Wright Field in Flight Testing. When I went there, there were six of us. This was in 1939. Six of us were doing experimental flying of aircraft and we were very, very busy. What we would do is go out and make an evaluation of the airplane. It was a pretty exciting job because you had to get good results. You couldn't just go out there and say, "It's a good airplane." You had to come down with hard figures.

The first exciting experimental aircraft I flew was the BL-239. It was a plane with the engine behind the pilot. The pilot would fly with a glass shaft between his legs. It had the propeller up front and 37mm cannon that would fire through the propeller hood.

Larry Bell was going around, bragging that he had the first 400-mile an hour fighter aircraft. He almost convinced half of our outfit that was true. When it came time to flight test, I was chosen to go up and do the flight-test myself. I knew I had to meet 400 mph or they were going to look down their noses, "What's wrong with him?"

When I got to the airport, the plane had all the gun ports taped over. This was supposed to be a combat aircraft, so I made them take all that junk out. It had to be in combat condition before I would test it. I ended up getting 362 miles per hour, not 400. I was very careful at high speed because it had an automatic flap on the radiator. As the temperature increased, the flap opened a little more. When it decreased, it closed. It took a long time of level flying, low power, for things to stabilize and that flap wasn't coming up and going back. When we sent word to Washington, Larry Bell went screaming to Washington. "What's wrong with these guys?" So Washington sent another pilot in to make an evaluation. He got one mile per hour more than I did.

The Bell FM-1 was a twin-engine fighter, elephant engine, and it had a propeller in the back of the engine, pusher-type. The reason was they wanted to leave a cell on each side for a cannon and a gunner - - a person sitting ahead of you in a cell, shooting the cannon. All the systems were completely electric. It was the first plane to have everything run by electricity, including the landing gear, propeller, flaps, governor and instruments.

One day I was asked to run a test flight with the electrical engineer at Wright Field. In development, they didn't balance the radiator for the coolant and the oil. In real cold weather—and this was a real cool day—you had to warm up the engine, because the oil would not warm up that fast. Then you'd shut off the engine, let the heat from the engine soak into the oil, and warm it up before you could take off. So, while we were doing this, I was sitting in there talking to the engineer about the airplane. He was bragging about how wonderful it was because everything was run by electricity. I told him otherwise, that it was a lousy airplane from a pilot's standpoint. If you were flying and lost the generator, you had only 30 minutes left on the battery to operate all the systems. If you didn't decide early to feather the propellers and jump out—and you laid on after you used up all the energy—you couldn't feather the propellers, so you had to land the airplane. So, you see, he was not happy with me not being happy with his airplane and the electrical system.

After we got the oil heated up, we taxied on the runway and immediately after the retractable gear went up, smoke started appearing in the cockpit—we hadn't even left the ground yet. I looked around to see where the smoke was coming from and he was out of his seat, dancing a jig trying to avoid a hot electrical wire. The wire would bounce around and every time it hit metal, would arc like a torch—arc welding. By then we were in the air. So, I wheeled around and landed. Fortunately, I was able to get the gear down— we still had enough battery. I stopped short of the ramp and shut everything off. I turned around as he was getting off the airplane and said, "Maybe we should think about this airplane." He was shaking his head. I never saw him on a flight again after that.

# Test Piloting, Beyond

## Najeeb "Jeeb" Halaby, Jr.

Najeeb Halaby, Jr. was a pioneer test pilot, a businessman, and flight administrator, whose influence on flight stands out. As a test pilot, he made the first transcontinental jet flight across the U.S. He served as Chairman of Pan American World Airways. He is otherwise known as the father of Queen Noor of Jordan. Appointed by President John F. Kennedy to head of the FAA, he remembers the experience fondly. He says, "The main thing about JFK is that let me run it. He had confidence in me." He would later head Pan-American Airlines for two years.

His remarkable career, in a sense, got a start in 1927. Charles A. Lindbergh was on nation-wide tour, and his father took him down to see the celebrated flier. Later, when Najeeb moved to California, he connected with Uncle Harry Willis out there. Harry soon convinced his mother she should buy her son flight lessons. There were jaunts to the airport to watch early pilots in their small planes take off, capped in 1932, when his uncle, Harry, took him to the air races where he saw pilots Roscoe Turner and Jimmy Doolittle. One extraordinary flier he witnessed would fly low, dip and pick up a handkerchief with the wingtip. By now, this young man was excited, hooked. Najeeb attended Stanford and did a little flying there, and then I went to Yale Law School.

As a professional test pilot, he set records and, on occasion, would even fly with WWII Aces, like Guadalcanal Ace Marion Carl, sent over to learn basics such as stall characteristics. He even showed Charles Lindbergh some analytics.

Najeeb Halaby, a man of the world, died in 2003, having made innumerable contributions to flight.

*Tell me about your mother buying you flight time at the beginning.*

My mother bought ten hours of flying lessons at ten-dollars an hour, and we flew in an old Travel Air. The instructor was a friend of Uncle Harry and gave me six-and-one-half hours of flying time and soloed me without doing a stall or a spin or anything required today. I soloed on a Saturday and came back on Sunday morning to take my seventh hour, but the airplane had been repossessed. The instructor apparently ran out of money. This was 1933, in heart of the Depression. I remember the plane had a big red sign on the propeller: "Repossessed" (Laughs). Uncle Harry was angry with him for not staying solvent.

*What was Uncle Harry like?*

Frankly, Uncle Harry Wilkins was really frustrated at not being a pilot himself. He was, in a sense, living through my flying. He suffered at not being able to pass the physical. He took me to watch the planes fly at LA International in 1932. Very few flew then, but I saw Will Rogers and Jimmy Stewart. There were very few private pilots. Western Air Express was the only airline.

In any case, I went to college and flew a little there. Then I went to Germany for a summer trip, and I was terrified by the Nazis. I decided then I was going to become an aviator, so I took the Civilian Pilot Training Program. I taught for the Army Air Corps and then transferred to Lockheed as a test pilot on P-38, Hudson, Lodestar and Ventura.

The chief pilot sent me up to an airfield, which is Edwards AFB now. My job was to test a cannon added in the nose of the P-38, to see if the shell casings would eject without damaging the tail.

*What happened?*

We had a wing-mounted camera and they fell away from the tail and didn't hit it. On the way back to Burbank, I was feeling pretty good, so I buzzed a tower at an airport. I came by fairly low, about ten feet below the top. I then pulled up and did a half-loop, and rolled out on top and went back to Burbank. When I landed, there was an Army Air Corps officer waiting for me. It seems that the Lieutenant General in charge of safety in the Air Corps was in the tower.

*Talk about an odd coincidence.*

Yes. He grounded me and took me off the contract with Lockheed. I got shipped over to the Navy on a contract. They were building a Lockheed Venture bomber for the Navy. I was to deliver this Lodestar to Commander Paul Ramsey at the Navy Flight Center at Anacosta, right across the river from National Airport. I delivered it and checked the commander out in it. He said, "What's a fellow like you doing in civilian clothes? You ought to be in the Navy." I said, "I'd love it.' This was 1942. He said, "Go down to the Bureau of Aeronautics tomorrow morning and sign up." I said, "Are you serious?" He said, "We need you in Navy Flight Test." I went down there, and a nice guy welcomed me and said, "We have a request for you from Commander Ramsey. Go up to Bethesda and take a physical, and you'll be in the Navy." I went up and passed the physical and became a Lt. Junior Grade.

*What was it like at Lockheed?*

There were three pilots at Lockheed. We were all lawyers, over-educated bastards. We had a group called "The Legal Eagles." Our ID card was the two-dollar bill. Around Lockheed-Fairbanks, they used to call us "Pepper Tree Heroes," because the pilots' shop was by the pepper trees. All the people there thought we were just overpaid, under-worked guys (Laughs).The people there did crazy stuff. If a pilot was sleeping between flights, sometimes a guy would slip up and light a match and put it in the sole of his shoe, and it would wake somebody up good.

*Did that sort of match game result any fights?*

No, we were a bunch of kids there.

*You said you were terrified on trip to Germany.*

I think the most impressive thing was the youngsters. I was in Nuremberg and saw them marching, and they were fanatics. That was in 1939. I was in law school, and that made an internationalist of me. I came back and joined William Allen White's The Committee to Defend America by Aiding the Allies. I gave a lot of speeches. The America First Committee was Lindbergh and he advocated staying out of the war. I was just the opposite side. Ironically, I was giving speeches against my old hero.

*Did that make you feel funny?*

Lindbergh had been to Germany earlier and came away with a different attitude. I went and got the feeling that unless we did something to defend the United States that they would be coming after us. After that visit, I came back to the U.S. on the last voyage of the Queen Mary.

*Any tough planes to test?*

The Hudson was a version of the Lodestar. They put a bigger engine on each side and it became a bomber for the British. Then they advanced to the Ventura and it had a first 2800 Pratt-Whitney engine. The main character these planes had to do was put more tail stabilizers and more rudders. It was hard to take them down the runway. You had to differentiate the engines. You would have to put more power on the left side and less power on the right side.

Early on, they had actor's agents and the like who flew as co-pilots. They tried to avoid the draft by being Lockheed co-pilots. I had a day delivery of a Ventura to the short field at Long Beach Naval Air Station. I had this guy as co-pilot and told him we wanted to drop the wing tanks. We had to drop them in a certain place, otherwise, there wouldn't be enough room to land. This was a delivery flight. I said to the co-pilot to put the flaps down—only thing he had to do. I would put the gear down and he, the flaps, because I was flying with both hands. We approached the field, and I put the gear down, and we got lower and lower and I said, "Drop 'em." But he thought I said to drop the drop tanks. I couldn't have been clearer. Just as we were coming down, the two drop tanks rotated off the holder, banged around the tail of the airplane. It pretty much made a mess of the plane. Do you know the moral of the story? Don't hire a draft dodger to be co-pilot and don't let them drop anything (Laughs).

*Any forced landings?*

Once or twice, yes. At Lockheed, I was up on a test flight and couldn't get the flaps to work. The problem is we had two big drop tanks. A lot of planes then had big drop tanks to extend the range. I notified and the chief test pilot flew under the plane in another P-38 and observed the flaps, and why they were not disengaging. So I rolled the airplane and I pulled several G's out of a dive. We thought that might bring the flaps down, but it didn't. They prepared the runway there at Lockheed-Fairbanks, so I came in. This occurred at lunchtime so thirty-thousand employees watched all this. It turned out someone had installed a valve backward and so the flaps and drop tanks wouldn't come off.

We had several P-59s there. We couldn't get the kind of speed out of them we wanted, so we decided we'd try to get an altitude record. I was instructed to get it to go as high as I could and to perform a dead-stick landing. This is 1944. They figured I could set the record. But I would have to affect the generator by flying fast enough, so the intake pressure would rotate the generator. Unfortunately, that didn't work. I had a time landing dead-stick. I finally cranked the landing gear down and landed without flaps. I think I got the landing gear down on final turn into the airport. I had dropped from 46,600 feet and, when I landed, I didn't have any fuel, so they towed me back to the hangar.

*But you did set the record.*

Yes, but I had to keep it a secret, because the Japanese didn't know we had jets. But it didn't last long (Laughs).

*Another story?*

Charles Lindbergh was chosen to fly this then-secret Air Corps jet, the P-59. He was in Washington, D.C., and so the military sent him down to the field. I had experienced the most hours on the airplane, so they had me check him out on the plane.

*What was that like?*

He was a very quiet, wonderful man. He was just wrong about the war. He was a great pilot. No question he could fly the airplane. He took instruction sincerely.

*Did you talk about the earlier difference of opinion? What did he say?*

He said, “We were on opposite sides on that one, weren’t we?” I said, “Yes, we were” (Laughs).

# Fire Pilot

## Kenny Roth

Kenny Roth was born in Kansas in 1924 and lived there until the fifth grade, when his family moved to Wilder, Idaho, a small town just outside of Boise. His first summer out of high school, he got a job as a fireguard for the Forest Service. Later on, after military service in the Navy, he returned home. He worked on his stepfather's farm and was out hoeing beets, when a friend appeared and asked him to go smoke jumping. When he and his friend found the jobs had been filled at McCall, Idaho, they tried Missoula. Nothing there. But it turned out two people didn't show at the McCall location. Kenny and his friend were hired. Thus began a long, somewhat circuitous career of fighting fires, first as a smoke jumper on the ground, and then in aircraft, as a fire pilot.

Roth worked for five years as a smoke jumper in McCall, Idaho. Then, after the fire season was over, he would go on the road crew. A friend talked him into taking up flying on the GI Bill. Roth got his private license, then a commercial license. He was logging by then and used the plane to fly back and forth around the South Fork of the Salmon River.

One of the workers at the South Fork had been flying with Johnson Flying Service out of Missoula, which was noted for its aerial firefighting, and he was due to take a leave of absence. He asked Kenny on the street one day if he would be interested in flying for him. Roth had seen that airplane flying over on its way to fires, as he sweated on the ground with his chainsaw. The thought of doing that type of work had clearly crossed his mind before. The year was 1952. Roth flew as a fire pilot until 1984.

*How did you get into the fire-fighting business?*

I originally started out as a smoke jumper. Right after the war in 1946, a friend of mine who was going to school up there in Moscow, Idaho, came down home and he had heard about smoke jumping. I had never heard about it. Anyway, he came back with this smoke jumping idea, but neither of us knew what it was about. He wanted to know if I would go up there with him. I asked him what it amounted too, and he said, "I don't know. Parachuting in on forest fires or something." I said, "That sounds good to me. Let's give it a whirl." I made fifty-six jumps and stayed with the Forest Service for five years. But I wasn't making any headway there, so I went into logging. In the meantime, I was learning to fly in the off-season down in Caldwell, Idaho, on the GI Bill. So that's how I got into the flying end of it. I started in 1952, flying.

*Do you remember your first flight carrying people out to the fire, and were you scared?*

Oh no. To start with, you had someone with you. I don't know how many trips they did, but I had an advantage. I was a squad leader when I was a jumper, and I sat up front with the pilot all the time we were going to and from fires. I had to work in the back, spot the jumpers, and toss out their cargo and whatnot. So I knew the job.

*What was the toughest thing you had to get used to when you started working as a jumper pilot?*

In a lot of situations, you don't have a lot of margin. I mean, it's a matter of knowing what you are up against. If you are flying up a canyon, you need to know if it's out-climbing you or not—or if you have enough room to turn around—things like that. You have to know what the winds are doing to you, or for you. If wind is coming over the ridge and you are flying too low, you're going to fly into down air and it will take you right into the mountain.

*Tell me one of the worst situations you've been in as a pilot.*

I've had a couple of close calls. I had a Ford Tri-Motor one time. We had lost another Tri-Motor that spring in a spray job accident, so we leased one from Orofino. When we got it we knew there was something wrong, but we didn't know what it was at the time. The plane just never performed like ours did. It ran all right, but we could only haul about half the load that we used to haul in our own. Of course, going out of Missoula here with the Ford, you always went out with full fuel tanks.

I had been flying over there in the Bob Marshall Wilderness Area, carrying a load of jumpers. I had six jumpers and a spotter. It turned out it was too windy for them to jump on the fire we went to, so we landed there at Spotted Bear, which is a strip out in the back country.

Then, when it started cooling down and getting later in the evening, we thought we either had to go out and jump or go back home. Taking off, the plane quit running just after I got over the timber. I was barely on top of the trees, so we ended up landing in the trees.

*You landed right on top of the trees.*

Yeah, the plane went to the ground though—the trees cushioned our fall. Luckily, we didn't hit something solid on the nose, so we got out of that all right. It turned out to be a carburetor problem. They had pulled some engines off some BT-15s, which were the same as BT-13s. They had a pressure carburetor. We weren't getting enough fuel to maintain the float level for the gravity feed. They changed a lot of things to get the engines turned out—mainly the pitch on the props, which were fixed pitch on the airplane. We figured that once I had ran about half of my fuel out, it lowered the pressure that much more, and that's what they figured was the reason for the accident.

*You mentioned a second situation.*

I had an experience in a DC-3 when I was spraying timber out of Bozeman, Montana. I got into some bad air as I was turning around and ended up going upside down. When I got it straightened out, I was headed into a rock cliff. The plane knocked the tail wheel off and also about half of one stabilizer, but I was able to get back into Bozeman with it.

*What happened specifically?*

I had made a pass pretty much down in the canyon, and I climbed out to turn around. It would have been my last load of the day anyway, because the air was starting to get kind of nasty, and the Gallatin River—with rock cliffs along the side. I figure the currents were the cause: the cool water and the hot cliffs, which were forcing the air that way.

They had our water tanks sitting up off the floor. Because the DC-3 is a low-wing airplane, that puts all that weight up about two or three feet above the floor. In other words, it raised the center of gravity up quite a ways. I think that's what rocked me over into an almost vertical bank. I went right over and back up. Then I was in the canyon—up in the air a ways. Then I lost some altitude and, when I came out of it, I was still probably four or five hundred feet above the river, but I was right over the edge of the rock cliff there. When I went by the cliff, I was pretty much standing the plane on one wing. That's when I hit the tail wheel and knocked it clear off the plane. I never did find it. It also tore about half of the elevator off and buggered up the stabilizer pretty good.

*Dropping too fast is hard on the airplane.*

You run into that situation every so often. They found out the hard way with B-52s how hard it was. You have all that weight hanging on those wings. You release it all at once, and the airplane pops the other way. The airplane was not designed for it in the first place. They lost four of them one year down in Boise—just knocked the wings off them.

*That must have been a shock, wasn't it, losing those men?*

As a pilot, you don't think about that. It's just like people having wrecks in cars. You go get into your car and go, and don't think anything about it. That's the way it is with flying—it's just something you do. That negative stuff never enters your mind. If it did, you wouldn't be flying.

*Is there any special problem when a fire crowns out? It takes the oxygen from the air*

.

No, you get into some turbulence around a hot fire, but that's because of the currents created by the hot air going up. It can get pretty rough when it really gets hot on the front of a fire. Normally you are not dropping slurry there because it's not doing you any good. You would just be wasting it. The time to fight a fire is when it is just laying down. Once it gets rolling, there is nothing you can do with a forest fire, until it slows on its own.

The life of a jumper is a good life. Those guys stay in good shape all the time. They have had some pretty good injuries for the time they have done or the number of jumps—also the terrain that they jump in. But they really do have a pretty good safety record.

*Did you see the Steven Spielberg movie,* Always*?*

I didn't see it, but my daughter did. I used to fly the B-26s. The fact is they used the one belonged to Johnson Flying Service before Lynch bought it. I flew it for Johnson and Lynch both. I flew it for Lynch for about four years.

I've heard in the movie they dropped retardant out of one plane on another, but that wouldn't happen. It would knock them out of the air—to start with. It would be the best way in the world to make sure you killed them.

*Ever have any unusual cargo?*

We dropped a lot of bridge timbers—timbers for lookout towers. We had timbers as long as a DC-3. It was quite a job. They would be heavy and would take three or four guys in the back end of the plane.

They would take both doors off and we had a chute built up to fit in the door there on the floor. It had a block on it so the thing wouldn't slip back because you had to be careful you didn't hit the tail with it. I also flew horses into the back country in the Ford. The vet would give them something to put them under. Then we would force them onto the plane. We had two of them in the Tri-Motor at one time. We hauled three in there to Big Prairie, which is in the Bob Marshall Area.

*Any trouble getting them out?*

No, it wasn't getting them out; it was getting them in because everything was going uphill, going in. Coming out they were downhill, you might say. They were saddle horses—probably up to 1,000 pounds apiece. We put them on canvas, tied their feet and slid them in.

*Was there one plane that was harder to fly than the others?*

I suppose it was the Ford Tri-Motor. It took more flying than any other airplane I ever flew—to really fly it and do what you wanted to do with it. It didn't have the margin in speed, for one thing. You were working within ten or fifteen miles per hour between stall and flying speed. That meant that working in the mountains you had a lot of worries. You had more problems with it than you would a faster airplane. But if you're talking about landing on short fields, it was good for that. It would get in and out of pretty short fields. The DC-3 worked real well, too, but of course, it wouldn't get in and out where a Ford would.

The engines on the Tri-Motor had all been changed when I flew them. There was no problem. They had different sizes of Ford, which a lot of people didn't realize. The little one, the Ford Four came out with a Wright engine, rated originally at about 220 horses, which they then changed over to a more modern Wright engine, rated at 360 horsepower. The big Ford had the Pratt-Whitney 1340s—they called them. They turned a long prop with slow rpm—to give them horsepower. Then they converted it to a Pratt-Whitney 985. They didn't work so well because they had to put a different prop on them. Since the engine wasn't really built for that prop, you had to turn them faster to get horsepower.

*What is your height going over a fire?*

If you were flying retardant, that's the only time you were really involved with the fire. Dropping jumpers, you didn't have anything to do with it. Of course, you always dropped the jumpers someplace where you were sure they weren't going to land in the fire. They wanted you about one-thousand feet above the terrain most of the time. For jumpers, now, I think they have increased that upward because they have different parachutes, which give them more control. It really all depended on the terrain and wind conditions. If you got a little wind, or they thought the drift was a little too great, they might come down lower, but they never got below a thousand feet to drop jumpers. On the other hand, in dropping cargo we used to come in low, right on the treetops. But they've changed that too, today. They want you higher, and they've redesigned the chutes so they drop higher. They're not as accurate as they used to be with cargo, because the higher you are, the harder it is to hit the spot with the cargo. Whichever direction the chutes happen to be going when they open, that has a lot to do with where they are going to land.

*Talk about an early flight.*

I flew the first slurry that was flown in this region with a Tri-Motor. It was almost like dropping cargo. You would open your tanks rather than kicking it out the door. When you are dropping cargo, you have someone else back there booting it out. The first planes we had we just had a hand lever in the cockpit. You'd just pull the lever with a cable on it and open the door. The problem was we had no way of closing the doors once you dropped. The authorities didn't want us flying back across the country that way because there would always be a little bit of stuff dripping. So they had to redesign the doors to close. Then they went to hydraulics. They didn't want you flying around dripping retardant.

The first retardant was borate. That's where they got the expression "Borate Bombers." It was the stuff they used, which was derived from borax. It was a real fire retardant, but it was also a soil sterilant, which they didn't realize at first, so they quit using it because of that.

*What's the most deceptive fire you've been in?*

I remember one for certain, because one of the jumpers got injured. They dropped two guys on this little fire, and one of them got injured pretty bad. The other was trying to take care of him, and meanwhile the fire was getting away. It turned into about a 50,000-acre fire before it was all over with. There were fire jumpers who got in, but they were too late. There was no way anybody was going to control it at that point.

*What makes a good fire pilot?*

I think just understanding the job and working at it. I know we have had good pilots in the past that came in, but they basically didn't like the kind of work. They didn't like that kind of flying; they didn't like being that close to fires. I remember one who was a real good airline-type pilot that worked for us when Johnson was doing charter work. He came here to fly jumpers. We were kind of shorthanded. He went out one morning to drop jumpers and cargo over there. I think we were in the South Fork, in the Bob Marshall Wilderness Area. I came back and left on another flight. So he went out with a load of cargo. He was dropping it, and the spotter went up to tell him he was going to have to slow down. He was tearing the chutes off the cargo because it was coming out too fast. He said, "How slow do I have to be—down around eighty?" He said, "The only time I get it that slow is when it's parked in front of the hangar." He came back and quit—right there.

*Was there one canyon that was especially tough to fly?*

There are some that can be a little narrow—or something like that. There are a lot of places where you have to fly down-canyon and drop and climb all the way back out. I did that over there on that Corn Creek Fire on the Salmon River. I was fifteen minutes between drops because you would have to climb so high to get back up on top and then drop all the way back down to drop. It was really hot down there. You have to baby your engines a little, so you don't burn them out climbing out. I shouldn't put it that way. You wouldn't just have one burn out and quit like that—I don't think. But if you are heating up for too long a time, you are going to do damage, and they will give up sooner than they would normally. That was one thing about Johnson Flying Service. They changed engines regularly. I think we only ran a thousand hours on DC-3s, which is way under what even the airlines would allow. The hardest thing about it was the power off and the power on. You couldn't really ease into it in a lot of situations. You would cool them down going down into a canyon, then pour the coals to it, climbing back out. And that was continual.

*It's pretty important what you've done—what the fire pilot does?*

I think so. I think jumpers are a big asset to the fire-fighting force. I sometimes think that there is a little bit too much politics in a lot of the fire-fighting divisions. They could use jumpers and retardant to a little better advantage, but it's a matter of one guy having authority over this county and one over another. You just don't step on one another's toes. I guess that's what it boils down to.

*Did you get involved in those politics?*

Not on my own, but I've been in a situation where I could see it clearly. I was flying out of McCall, Idaho—I was based there, flying jumpers for a while. This was out on the Payette Forest, Region Four— this is Region One.

Anyway, this happened on the Nez Perce, which is Region One, right across the Salmon River. That was the time between their operation and ours. A thunderstorm went through, and we went out with a load of jumpers just looking for fires. We'd drop jumpers, if we spotted any. Well we spotted one just across the river. We called it in. We told them we were sitting there with sixteen jumpers, and no place to put them. We said we'd be glad to drop them on their fire for them. They said, "No we'll take care of it." They did—2,500 acres later!

The sixteen jumpers could have gotten it at that time. That was my feeling and the feeling of everyone else in the airplane. But I guess that's neither here nor there. Maybe we wouldn't have got it either. I guess it took them at least an hour before they got their people there.

*Your worst landing was?*

I don't know. There were some rough fields; there was one called Crow Foot that was short and steep. It was located on the Salmon River. The fact is if it wasn't for being so steep, it would be too short to land on. It was like landing on a steep hillside, going up a hill. Taking off you just opened up the throttle, and you just sort of fell off of there. It wasn't all that difficult, but it was one of those places where you'd have to have good air when you were landing. You'd be coming off the river and making a ninety-degree turn, then landing. The biggest problem was going too slow, instead of too fast, which is a main problem you'd have in landing on a short field.

This was short so you would have a tendency to slow your airplane down, but that's where you'd get into trouble. You had to have the power to carry you up. Even after you got on the runway, you would have to use power to get to the top. Also, it was too steep to turn around on. You would have probably turned the plane over. We used a Travel Air in there. But this is all stuff you get used to. It's not like I'm trying to make a big deal out of it. Anyone who is a good pilot can learn to do that, if he wants to. But you will run into some people that absolutely won't do it—and they shouldn't.

*What has flight meant to you over the years?*

I've met a lot of good friends flying all over the country. Of course, I don't get around that much anymore, but it always seemed if I got around an airport someplace, I would run into somebody I knew.

*It kind of brings you together.*

Yeah, I think that's one of the big things. The only big money in flying is if you own the operation. Some people make good money in the spraying business. Other than that, your big airline jobs are the only ones that really make good money. I make more money now driving a truck than I ever did flying.

*Do you sort of miss flying now?*

Oh, yes. I enjoyed it when I was doing it. I was doing other flying through the winter. I was getting too old to do that kind of flying. Sixty is all they will allow you. I was flying tankers early in the season down south and coming back here and dropping jumpers. Then I wasn't doing anything one winter, and I got to driving trucks. I thought I'd just as well stay here rather than looking for a job for the next four or five years. So that's kind of how I got out of it.

*Isn't it a little easier on the ground?*

No, I don't think so. For one thing being in this job there are about eighty drivers where I work. We run those trucks around the clock during the week. Of course, being down at the bottom of the list, I drive at night, and that gets old. Tonight, I'll probably go to work around four or five o'clock in the afternoon and go till three or four in the morning.

*What do you think when you hear that there's a fire some place? Do you kind of suddenly wish you could be in an airplane?*

Yes, I'll have to admit that you do think about it. I still have to look up every time an airplane goes over to see what it is, unless I recognize the sound. It's just like jumping. I kind of missed jumping for a long time after I got out of it. I would be down there on a chainsaw, cutting logs or something, and I'd think, "Well, jeez, I should be up there in that airplane going to a fire instead."

*Did you know any old-timers who did spraying?*

I knew some. As far as I was concerned, there were probably as good spray pilots as there were in the country over in Lewiston, Idaho. They both died. They didn't get killed in a crash. I think they died from the stuff caused by chemicals they used for years.

For a long time, these two were spraying for Ivan Guston, another old-timer. They were flying a Stearman biplane out of a field in Lewiston. The field had slopes over slopes and, from one end to the other, you couldn't see the other end. One day one of them was taking off, the other landing. They locked horns—their left wings—right out there at the crest of the hill and just demolished those airplanes. Neither one of the guys got a scratch on them. I ran into their boss shortly after that down at the Shepp Ranch, Idaho. I told him, "I hear old Wayne Hughes and Don Shoemaker kind of tangled the other day." "Yeah, them goddamn fools." He really liked them, you know, but he was a little bit upset with them wrecking his planes. But he didn't fire them or anything.

*Who was the most memorable character you ever met out there?*

There were a lot of good pilots and a lot of good people. There were a lot of old-timers that I knew that were getting out of it about the time I was getting into it. Like old Penn Storrs. He was a kind of legend in his own time. He got killed in a Ford Tri-Motor over there spraying sagebrush between Townsend and Helena.

He was one of the early mountain pilots, you might say. Of course, he flew a lot out of McCall. They were hauling supplies and mail into the back country all winter long. He also did a lot of ski flying. Then during World War II, there was a B-25 bomber that crashed and landed on Loon Lake. The people all survived—they actually didn't get hurt. They landed on that lake, but they ran up into the timber. Some of them decided to walk out to see if they could get help. Penn Storrs happened to be flying across there one day, and he spotted the wreckage. This was a week or so afterwards. He landed in there. He had skis on his airplane—a Travel Air—and picked them up. Of course, he got in hot water over there with the Air Force, because they figured he should have notified them before he did it. But he figured they had been there long enough. He didn't ask anyone—he just went in. They kind of raked him over the coals, and then gave him a medal for doing it.

# Experiencing the Battle of Midway from the Water

## George Gay

George Gay's grandmother came across the prairies in a covered wagon. Young George listened to her stories of her adventures. Later on, she would be the first one to take him up in an airplane. That took place at the Dallas State Fair, after his father had refused to go near the plane and his mother had said, "Absolutely not." At that point, his grandmother informed him, "I'm not scared of that thing. Come on." When they landed and were walking back to where the crowd had gathered, the old woman said, "You know there was a time I could have used that thing."

Later on, at Texas A and M, studying engineering, news reports told of the rumblings of Hitler in Europe. George decided he'd rather be a flier, if he could. He went into the Army Air Corps, which leads this patriot to a story about how he almost didn't make it into the military. We start the interview, and Ensign Gay, who witnessed the greatest naval battle in history, at the Battle of Midway, tells how he almost didn't get the opportunity to serve his country.

I went to Randolph Field in San Antonio. They told me to get there early in the morning without breakfast, which I did. They then gave me a slow going over. Then we broke for lunch. I went to the PX and got a cheeseburger, malted milk and a Coke, came back, and they slapped me down for the heart test. If I had known they were going to give me a heart test, I would have gone somewhere and laid down for that whole hour. No lunch or anything. Anyway, after they slapped me down for a heart test, the man said, "We can't use you. You will never be a pilot. Go home and forget it." I flew five years with the Navy and 30 years with TWA and I think I made a liar out of him. (I learned later the best way to pass a Schneider test was to have a hell of a good hangover.) I'm sure that the doctor that came up with this test had never talked to a pilot or seen an airplane, because it didn't have a damned thing to do with flying an airplane.

They gave me the Schneider test after I had been asleep all night in the Rice Hotel down in Houston and I evidently passed it. The guy said he saw a little glitch in my EKG, which I have always had, but they didn't worry about it, so I got into the Navy.

*You were the only survivor of your squadron. Could you talk about that? What was the plan?*

Well, the plan was rather loose. The problem turns out to be logistics. We were in antiquated airplanes that couldn't go to the altitude of the dive bombers and the fighters. I never have been able to understand how the American people expect the military to defend this wonderful country with two kinds of equipment: experimental and obsolete.

Anyway, we had old airplanes and the carrier was not anywhere like what they have today. Some of the torpedo planes had to be down on the hangar deck. But you couldn't put the whole air group on the flight deck, so the fighters took off, then the dive bombers. This meant that it was 18 minutes later before all the torpedo planes could get into the air and join up. My skipper told me, as navigation officer, that he was not going to go where everybody else was going because he thought they were going to the wrong place. He said that he was going to cut across and try to make up these 18 minutes. Hopefully, he figured the other would go out and find they were in the wrong place and turn and come to where the Japanese were. If we all got there together—the dive-bombers and the fighters—we would be together and coordinated, but the problem was the torpedo bombers got there first. When we arrived in the torpedo bombers, we pulled down their fighters. I found out after the war that there were 75 Zeros in that combat air patrol, and they all came down to get us. It didn't take the Zeros long to wipe out our fifteen old, beat-up planes. All we had were 30-caliber machine guns for defense and that's about like blowing a wet match through a straw. On the other hand, the Zeros had twenty-millimeter cannons, so they just annihilated us right off the bat.

*When was your first realization you were trapped?*

It didn't take very long. What happened was that we saw some smoke over there. By the time we got close enough to figure out what was really going on, it looked like the battle was going. We were late so we started in, and by the time we got in close enough to see that they were landing airplanes, which was the thing we couldn't tolerate, the man who gave the attack order. I was Tail End Charlie in the squadron's navigation officer. The thing that was unbelievable was that I was sitting back where I could see the whole thing. It was all in front of me. Zeros were just swarming all over everything. They were pushing each other out of the way to get at us.

*How long did it take?*

It probably didn't take as long for them to kill everybody as it does to tell about it. My squadron was annihilated. I'm the only one that got in close enough to drop a torpedo. That presented a problem because I'd never carried a torpedo before and I had never seen it done before.

*How long before you dropped the torpedo and when you and your plane went down?*

That wasn't very long either. Actually, I'm not positive that I dropped my torpedo. The electrical was all shot up and when that didn't work, I pulled a cable's manual release. I think I lost 2,000 pounds of torpedo weight. I know I was bouncing around in that anti-aircraft so bad. I know that I never have claimed to have gotten rid of my torpedoes, much less gotten a hit.

If the torpedo went, as it should and hit the Japanese carrier, it might not have exploded because the detonators were no good. Our equipment as I said was absolutely lousy. As long as it took me to fly about five miles, I went clear through the Japanese Navy and out on the other side. Then a Zero shot me down. When I was shot down, I was right in the path of the whole Japanese fleet.

*How did you feel sitting in the water watching the great sea battle?*

Well, Winston Churchill called it a "fish-eye view." I was actually scared to death. Of course, I was curious too, trying to observe whatever was going on. I didn't have to look in too many directions because the whole thing was right in front of me.

*You said somewhere that the planes coming into the water looked like pieces of orange peel.*

Well, that's kind of a reference to when I was younger. I remember going around in Detroit, eating an orange. When I would throw the peel off and you could see it until it hits the water and it would just appear to disappear. With the airplanes, I'd see them until they hit the water. You think of some of the darndest things right in the middle of a mess like that, and that's what I was referring too with the orange peel.

I had a raft, which came out of the airplane while I was trying to help my gunner. But he was already dead. The raft had just come out of the airplane. I didn't know how. This occurred about nine-thirty in the morning, and the Japanese carriers were hit at about that time. I remember they made a wide turn and came back and stopped right where I was. I just sat there in the water in the middle of the Japanese navy all day long, watching those carriers burning and blowing up. I watched their cruisers and destroyers going around picking up pilots that were going into the ocean. I was afraid they were going to pick me up. Fortunately, they didn't.

I had some debris out of my airplane to hide under and they came by and saw it a lot of times, but they didn't know there was anybody there, so they didn't stop. During the night, most of their heavy stuff left the area, except for two big cruisers who tried to tow the carriers away that were not burning so badly. One carrier was burning too badly and they couldn't even get close. With the other two, they tied towlines. The cruisers were getting kind of nervous, though. From where I was, I heard those lines breaking.

They didn't want us to have any of the ships, so with two of them, they put crew aboard to scuttle them. With another, I watched cruisers shooting them at the water line. I guess a sub was nearby and put some torpedoes in it, too. All three of those sank during the night.

The next morning the PBY came by and saw me, and he came back late in the afternoon and picked me up. I figure I was in the water, give or take a few minutes, about 30 hours.

*What images or sounds stick in your mind?*

Well, I think of the strafing. As long as my airplane was visible, the Zeros were strafing me. The bullets were flaking around in the water. But when the airplane sank, at that point I think, I realized that the dive bombers were coming down. Then they left me and started getting back up where they belonged. But the gunfire and explosives I've always attributed to running the sharks off. I didn't see any shark activity at all. I think those tremendous explosions that were shaking my guts so badly just ran them away.

*They were literally shaking your gut?*

Oh gosh, yeah. When that stuff would go, I would turn my back to it, so that it wouldn't hit me in the belly. I could take it on my back better than I could on my stomach.

*Do you remember the mood of the pilots when you first took off?*

I've always referred to it as like something before the big game: you know the thing is going to be hairy. In this case, we knew we were outnumbered; we knew we had old, obsolete equipment, etc. However, as the Commander put it, we didn't have any choice. We had to go and defend our country. I think nearly everybody considered that there were going to be people that wouldn't make it. Most just figured it won't be me.

After the whole thing was over, I came back to the States. I felt responsible for talking to all of the relatives that I could, and nearly all of the mothers and wives. They always expressed that they felt like somebody might have gotten it, but they figured their people wouldn't. My mother felt that way, too. In my estimation, that doesn't justify her optimism. It just happened to turn out that way.

*Do you remember the most touching experience of talking to relatives?*

I did what I did, though under normal circumstances, that's the commanding officer's prerogative. He is supposed to write letter and at least, if there is a body to be sent home, see that a good friend goes with it. Unfortunately, there was nobody to do it except me. I think the most memorable thing was when one man told me that he'd rather I didn't come around anymore because every time his wife saw me, she got sick.

She felt her son should have been there, instead of me. That was a little hard to take. But again, I tried to be understanding. I just quit bothering them. That happened in a place where I was going to school for Trans-World Airlines.

Those people I talked to I tried to explain what happened as far as each man, but I didn't get too specific. You get 29 guys killed right in front of you in a couple of minutes, you can't really go into details.

They were all just an ordinary bunch, as I said in my book. They were just like any other group. Whitey Moore was a little fellow, had white, fuzzy hair and was kind of a mascot for the squadron. Jimmy

Owens, who had been a football player in L.A., was quite a guy. Squires Evans was the brains of the whole outfit. A man named Teats; we called "Plywood" because he was from Oregon and told us how plywood was made. I remember he stood up in front of the room one day and said it had come to his attention that most of us were completely ignorant about how plywood was made. He spent a couple of hours explaining about how you make a plywood sheet. That started something.

I have run into some things that you just couldn't hardly believe unless you were there. For instance, I had a man write to me not too long ago and implore me to send a picture and some information to a friend's daughter, whom he said was eight years old. Well, I wasn't about to do that, because I did that before, and the father walked up to me at an airshow and said, "Just who in the hell do you think you are, approaching my daughter like that?" Another thing that is amazing to me is how convinced people can be that something is true, whether it is or not. For example, a guy came up to me at an airshow once and said he was in the Navy with my son. Well, I said, "That's interesting. I didn't know I had a son." He said, "Oh, yeah, he was in my squadron, talked about you all the time." I said, "Well, about 35 years ago, I would say maybe it was mine, but in these days, forget it. There is nobody that age that could be my son." Another fellow gave me hell. He was reading me the riot act and said I was the only instructor at Pensacola that had given him a "down" and that had damned near cost him his Navy career. I said I was never at Pensacola, much less an instructor.

*Why do people do that?*

I haven't got the faintest idea. I know one of the things my father always said. It is not what you don't know that gets you into trouble. It's what you are sure of that ain't so.

*Psychologists talk about "survivor guilt." Ever have any of that?*

I never have. It just happens to be one of those things that just came about. I can't explain it.

*Was there ever a time when you had to catch yourself and tell yourself you were a hero?*

I don't consider myself a hero. I happen to be a lucky survivor. I am somebody who got through a situation and got an amount of publicity because I was at a place. This happened at a time when the story was needed. I think anybody who goes into combat is entitled to some recognition, but usually they don't get it. I just happened to see the whole battle and at that point, it was the largest naval battle. Some historians have contended that the Battle of Midway was the most decisive battle in history, along with Stalingrad and Guadalcanal. It's supposed to be the biggest thing in the world. Well, I sat through the whole damned thing and saw it. Then I'm picked up and brought back to our Admiral, and I'm the first guy from our Navy that he talked to. He used the story I told. All this is so unusual I have to admit that sometimes even I don't think I would believe it, if it hadn't happened to me. I certainly didn't do as much as my 29 friends who lost their lives. They died fighting for their country. Now they are the heroes. I just happened to be lucky.

*What else was lucky about the incident?*

That morning the plane found me the ocean was completely flat. There wasn't a whitecap on it. If there had been any whitecaps, he would have never seen me. That's one of the things that is very unusual. Of course, the Pacific is famous for those kinds of things. But what is strange is that he not only found me once, he found me twice. I mentioned how, when I was shot down, I went to try and help my gunner because I thought he might be still alive and when I come up choking and coughing, I run into my life raft. Just one thing after another happened to save my life. As I've said, the Japanese kept writing my number down, and I kept trying to rub it out.

When they picked me up, they flew me into Midway. They put penicillin on my wounds and gave me some shots of liquor. The next day I was flown down to Pearl Harbor where the doctors operated on me. My problem at that point was dehydration. When anybody walked by me for a couple of days with a cup of water, I'd get hiccups. But once I began to take on liquids and eat, I came out of it fairly well. I had two bullet holes in me and a burned leg. I lost 30 pounds out there in 30 hours, but everything considered, I was in good shape.

*Did you always want to be a commercial airline pilot?*

I was on television on "Today" and Dave Garroway, the host, asked me when I got out of the Navy, what I wanted to do. I said, "Well, I would like to be an airline pilot." I got telegrams from every airline in the country. I picked TWA, and I have never been sorry about that. At TWA all my buddies in all my classes were Navy pilots, so my treatment wasn't that special.

*Any other strange episodes flying?*

One time I had a flight going from Paris to Athens to Cairo. When we got to Athens, we had mechanical problems. As a result, when we got to Cairo, there wasn't time for our legal rest to make our normal turnaround. What they did was put a Cairo crew on our airplane. They went up to about 12,000 feet and nobody figured out why, but they went into a dive and plowed into the desert. Everybody was wiped out.

We had an accident in Gander, New Zealand that was pretty hairy. It was right after the war and a snowstorm flew us right into the ground. We banged up the airplane pretty well. Everybody thought I was wiped out on that, but I wasn't.

*How many forced landings have you had?*

Actually, I haven't counted. There are many, many of them in the piston days. On the jets, I only had two problems, both of which were precautionary. We shut the engine down because of oil pressure. I always thought the main advantage of being in a jet was that the passenger can't see that the prop sitting out there is feathered and know the engine is dead. On a jet, you can shut it down and they don't even know it.

I was flying out of New York going through Cleveland on a regular basis. Three weeks in a row, within one minute of the outer marker at Cleveland, I lost an engine, but was able to go straight in. It got to the point that the Cleveland mechanic would say, "Oh no, not him again. We got an engine change."

I lost quite a few piston engines. Some of them were pretty bad. One of them I remember the fire was going clear back to the tail of the airplane. We managed to get down on the ground and get it out. Fortunately, with all this, I never damaged, or hurt, a passenger in my whole TWA career, or never had one injured.

*Any strange phenomenon you've seen while in the air?*

I think the most impressive thing I can remember is going across the North Atlantic and looking up and seeing a comet up there. I figured out just about, where it was in the sky. So I went to Europe, I come home a week later and I see this same damned comet. It looks like it's almost in the same place. That causes you to start thinking about how big it is out there in the universe and how this little speck of dust that we live on is so infinitesimal. It makes you wonder how important I happen to be in this universe.

*Any crazy passenger stories?*

I had to get up and go to the bathroom one time. This time this man stopped me and said, "We are going too fast. It's hurting my ears." I said, "Well, that's just the pressurization. Get some gum." He said, "Oh, no, I know what it is. It's speed. How fast are we going?"

I gave him my ground speed which was 190—or whatever it was. He said, "Boy, I sure wish you could do something to help me out."

Hours later, we are coming in to the coast of France, and I meet this guy again. He said, "Hey, boy, I sure do thank you for fixing my ears up. I appreciate your slowing down." I said, "Oh, well, that's all right." He said, "By the way, how fast are you going now?" I said, "Forty-five miles per hour." He said, "That's wonderful. My ears cleared up just as soon as you got back and took care of things for me." He said, "By the way, does that make us any later getting into Paris?" At that point I wondered how he was smart enough to buy a ticket, or whether someone might have bought him a one-way and sent him to Europe, hoping he'd never come back.

One time I was on one of the new flights that had a sleeper. It was like a Pullman. People would make their bunks down at night. I was going to the bathroom in the back again and this time two beautiful legs come popping out of the upper bunk on my left. This gal stepped out. I guess she thought she was at home or something, and she was completely nude. She hit the deck right in front of me, flat-footed, with her back to me. When she realized someone was there, she made one more move and was back in that bunk, zip. As I normally say, I never did see her again, but I sure did see her behind. You notice little things like that.

I remember a movie actress who was on one of those flights with a boyfriend of hers. She wanted me to sign something that gave her the authority to say that she belonged to the Mile-High Club. You know what that is?

*No.*

You have to have intercourse above a mile to be in the Mile High Club.

*What did you say to that?*

Well, I wrote out on a piece of paper and signed it for her. I cautioned that I had to take her word for the fact. I said I could acknowledge that she was on this flight.

*It's been a good life of flying, hasn't it?*

I always thought it beat working. By that I mean, I didn't have to fight the commuter traffic. You don't have to be crazy to fly, I guess, but it helps.

# The Submarine Chaser

## Bill Heath

Bill Heath enlisted in the Navy in 1939, and following boot camp was assigned to a WWI Destroyer, the *USS Clemson*. He ended up in Panama. After a transfer from the ship over to base radio in San Juan, Puerto Rico, he got a chance to fly as Radioman Third Class in a squadron that was flying PBY's. While there, he applied for flight training and, around Thanksgiving, 1941, received orders to report to Pensacola for flight school. Because the war had started, he completed one of the fastest flight courses ever given by anybody—instrument training, formation flying, navigation, and bombardier training, celestial and dead reckoning—in seven months. I asked him about being an enlisted pilot.

Being an enlisted pilot was an unusual thing. Once when I was flying into Alameda at that time, we didn't have hard hats. We would just wear our white hat or a baseball cap. I landed and the guy puts chocks under my wheels and I climb down wearing my white hat and dungarees. He stands there, looking all around, confused. I say, "Who are you looking for?" He says, "The pilot." I say, "I'm the pilot." He says, "No, I'm looking for the pilot." He wouldn't believe me, so I say, "He's in the backseat." (There was a bag of sand back there for ballast.) The guy then climbs up on the wing and looks in there. He finally realizes I am the pilot, so I say, "Okay, are you going to give me the yellow sheet to sign, so I can go get a cup of coffee." That sort of thing happened a lot.

We were in Port Lyautey, French Morocco, and North Africa and were flying down to Bathurst one night. I was navigating the flight. Our briefing officer told us to stay away from the Canary Islands, a Spanish territory, which was very hostile toward Americans. It was a long flight down—maybe ten hours, over water. No lights on the shore. About two a.m. I wanted to shoot a star fix with the octant. I grabbed one of the crew to record the time when I said "Mark." While I was getting the octant lined up on a star, I saw a streaking light passing through the lens. I asked the timekeeper if he had seen it. He said, "No." I tried again, and again I saw the streaking white light. Still the timekeeper didn't see it. All of a sudden, about half a mile to the rear, there was a blast that was obviously anti-aircraft fire. I realized those streaks I was seeing were tracer bullets. I told my comrade: "I know exactly where we are. Let's get out of here!" Of course, it had to be from the Canary Islands.

*How early did you get interested in flight?*

I can remember when I was a kid. I lived in Bayonne, New Jersey. I used to go to Carney. They built destroyers there. I always wanted to go in the Navy, maybe because my father was in the Navy in WWI. We'd go see those little old destroyers they were building, but I tried hard—barely maintained average. Still, the only time I ever skipped school was when I went to Newark Airport to see the GEEBEE.

After flight school, July 1942, the squadron I was assigned to was in Alameda, California, and was picked to be equipped with some top-secret equipment that would pick up submarines under water. There wasn't anything on the West Coast we could do with the equipment, so they sent us to the East Coast. Then they told us to pack our white uniform for about a two-month stay. This was around May of 1943. Guess what? We ended up in Iceland, and our duty was all secret stuff we couldn't even tell our families about, though we didn't go back home for two years.

From Iceland, we went to Wales. Flying down there, I saw a bit of action: Germans flying out of France and trying to intercept us in the Bay of Biscay. Our standard patrol would be out of Pembroke Dock down to landfall on Spain, back to a point off France we called "Junker's Junction." They called it this because of the JU-88s, which came out from the Brest Peninsula. This enabled a rather short trip to intercept our North-South route. On my fourth trip, which was a three-plane sweep, we were flying a rather loose formation, going down; however, going back the weather had turned sour, so we tightened the formation. When we were in the vicinity of "Junker's Junction," the plane on our right, which was the one closest to France, sent a coded message saying they were under air attack. We immediately dropped down under the clouds to join them. The clouds were breaking up and everything seemed quiet, but they had been shot down. We and a P-Boat stayed and searched the area, until lack of gasoline required us to leave. This was the first engagement in WWII of the U.S. Navy and the Luftwaffe. These days later, a British sea-going tug rescued the three crew from that plane. Overall, our guys were credited with four Junkers. Not bad for the old fighter.

Then they sent us to Africa where we flew the Straits of Gibraltar. That's where we sank three submarines. They were coming from the Atlantic into the Mediterranean. We just flew a fence—a two-and-one half mile fence between Africa and Spain.

*Any problem picking the subs up?*

No, every sub we picked up, we sank. Flying the fence, our highest elevation would be 100 feet above the water. We received a commendation for this because, during the year and one half, we did what we did without losing a single airplane in the straits. In other areas, we lost 12 aircraft and 35 crew members. I was in two of the lost aircraft.

*How'd that work?*

We had two planes out there all the time, but not at night. They sent over some blimps to do the honors at night. On one occasion, the weather people told us they expected fog in the morning. We got in our planes and stayed overnight in Gibraltar, then went out early in the morning. At about seven, we'd get to our station. We always had two planes out who'd try to stay 180 degrees apart. One of our airplanes picked up a signal a little after eight a.m. We also got the signal. We had a little button right on the yoke. We pressed that button and the smoke lights would be fired out of the tail hatch. We marked our contacts with float lights, two on initial contact and one on each individual contact, thus giving us a path to follow on the fourth contact. Thirty retro rockets would be fired off at the same speed the aircraft was going forward. An intervalometer would fire three rows of ten bombs.

We were among the first to have sonobuoys, which was a long tube of electronics with an antenna on it. When it hit the water, it would drop a hearing device that could detect underwater sound. Then we could tell if we got a hit or not. This time we had lost the location of the doggone sub three times. Then we'd pick them up again. The fourth time around, we dropped our bombs and the other plane dropped his. He got three hits. We got three hits. Then when the sub surfaced, the British came up and let them have it with the hedgehog rockets.

The first sub we sank we had the same thing happen. We'd lose them and then pick them up again. One time British ships were dropping those markers down there. The Americans told the British to get the hell out of there, so they could do their job. They backed off and the Americans went in and got a couple of hits. Next, the British went in and shot the sub. I wasn't there this time, but apparently the sub came up and when it did, the bow came up high (the stern was still kind of low) about that time, a British Corvette and the American PV-2 each made a pass on them. The poor bastards tried to get out of the submarine, not knowing they were already sunk. Then, the British ship blew off the tail of the other British Corvette.

*What about firsts?*

We were the only units in the world to have the equipment. I mentioned we were the first Navy squadron to get radar; the first to get sonobuoys; and the first Navy squadron to engage the Luftwaffe. Also, the first submarine we sunk was the first one ever sunk without first making a visual sighting.

*Did your sinking the subs affect things?*

The area admiral commended us by saying that VP-63 turned the Mediterranean into an allied lake. In fact, the Germans didn't sink another allied ship in the Med during the rest of the war. I know this saved many a GI who got to Italy from North Africa. They stopped sending subs through after that. We still kept our planes up there, though. Our story here: The French had earlier doubted our ability to hit subs in Gibraltar. As a consequence, we invited them to send one of their subs through. We knew about what time they were coming because we didn't want to think they were a German sub and actually sink them. I was flying that day and we picked them up. We only dropped a practice depth charge. Practice depth charges weren't very big, but they were equipped with a shotgun shell, so if they hit the sub, they made a big bank. We sure gave them a welcome. After that, one of our guys says, "Let's go down about ten miles and do it again." So we did.

*What's the funniest thing that happened when flying after the war?*

I was instructing students one night in Corpus Christi, Texas, back when flying saucers where a hot item, and I was sitting in the left seat, coaching a cadet through night landings when the tower directed us to leave the traffic pattern for about five minutes. We climbed to about 1,500 feet on a SE course, which took us toward the Gulf, which was very dark—no moon. The cadet was making a shallow turn when I saw a light. It appeared to be slightly below us and moving. I asked the cadet if he had seen it. He said he did. I told him to hold a steady course, which he did, but the light still seemed to be moving. I called the tower to confirm if he had any traffic in our area. He gave me a negative. Finally, I took the controls and went right for it. It was farther away than I thought, and now I was thinking UFO. We finally got to it and discovered it to be the gas burning off an oil well. We all got a good laugh at that one.

One time I had a close call, nothing I had imagined. I was flying with co-pilot Bill Craddock and Frank Bowen, navigator. We were loaded with gasoline and booze that we picked up in San Juan and Trinidad. Here it was about a week away from Christmas. It would soon be celebrating time. After a little while, we leveled off and headed in an east, northeast course. The sun began to settle in behind the horizon and darkness took over. The flight was going well in all respects. From time to time, the crew relieved each other on watch. Up in the cockpit, we were taking turns snoozing. Then, just about 2:30 a.m., all electrical power in the aircraft went out. We couldn't read the instruments because there were no lights. There was suddenly a very strong odor of gasoline. The plan's captain, Paul Ware, a citizen sailor of the highest quality, appeared between the pilot and co-pilot's seat. He handed us a flashlight and explained what was going on. At least now, we could see our instruments. There was a stream of high-test gasoline running down the bulkhead (wall) and pooling in the bilges, a major leak. It was obvious that something had to be done immediately or this plane would soon become a fireball.

Paul and Lt. Craddock agreed that we had to have a look in the section between the overhead wing and hull of the fuselage. This was where all the gasoline valves and connections were located. To do anything the individual had to stand on the navigator's table, exposing his body to the slipstream, remove a large panel and with one hand feel the valves and connections.

With a gun belt around my waist and it lashed by rope to the Nav table, I tackled the job. Bowen and Chaisson held ankles and as they did, Lt. Craddock dropped the airspeed down so that the slipstream was not as great as we had anticipated. Standing on the navigator's table, I could reach the port inspection panel over the main gas valves and strainer area. After unlocking about three dozen fasteners, I watched the slipstream rip the entire panel off, thus breaking out the port blister plexiglass. Working with one hand, by feel, I tightened all the gas line clamps that I could reach. This couldn't have taken more than five minutes, but it seemed like five hours. After dropping back inside the plane, I saw Bowen still mopping the overhead with a rag and heard him advise that it appeared the leak had stopped. We opened up the plane, fore and aft, to air it out. By this time, it smelled like a gas station.

Strangely enough, when meeting Paul Ware at a reunion about 5 years after this occurred, I had a strange recognition. Paul brought up the subject and I absolutely could not remember it happening. It's strange the way the mind handles traumatic shock. We made it back to Port Lyautey in one piece on December 23, 1944.

*What was your first assignment after the war?*

I was stationed in Norfolk, Virginia, attached to VRF-1 (Air Ferry Squadron No. 1) my wife was pregnant and about 3 weeks from delivery. I was to leave about 1 week before birth, drive up to Bayonne and take her to the Brooklyn Naval Hospital when the time came.

Back at the barn I was still living in the BOQ and at work was giving my share to Uncle Sam by proving the "Safe and Expeditious delivery of Naval Aircraft." That was our Motto. It was early November and my plans were to take leave shortly before Thanksgiving for two weeks and bring Clare and the baby down to Norfolk to an apartment I would rent.

Everything was going well until that one morning about the middle of November. As usual, I had my travel bags in my car and was sitting in the ready room minding my own business when this big Lieutenant says, "Hey Heath, you're qualified in the PBY-5, aren't you?" I had the feeling that he was telling me, not asking. He said, "We got a nice trip for you just as soon as we find you a co-pilot." That didn't take long because he grabbed the first guy who came through the door who was also a Lieutenant. I was still a Lieutenant Junior Grade. My new co-pilot's last name was Moore, a real nice guy, he didn't fuss at all. He said that he had seen a lot of PBY, but had never been in one. That didn't even bring a smile to the first Lieutenant's face. I finally asked him, where are we going and he said, "We have a P Boat that has just come out of major overhaul up in Philadelphia, and it is going to Seattle for storage." I gulped because I had only about two-and-one-half weeks to play with and flying the slowest aircraft in the Navy inventory. On top of that we had to take what was called the U route, which meant that we had to go down South to Pensacola, west to Corpus Christi, then to San Diego, then to Alameda, California, then into Seattle. After delivery, we would have to fly our way back to Norfolk. This situation reminds me of a flight officer I was to work with later. He had a sign on his desk that said "Bitch you may – But Go you Must," so away we went. I was delighted to see when we got to Philly that this plane captain, the enlisted engineer, was an old buddy from VP-63. They didn't come any better than old Woods. We decided to fly the P Boat back to Norfolk that afternoon and give Lt. Moore another night at home.

The weather was good the next day and we got an early start. The old slow boat got us to Pensacola where we spent the night. Then onto Corpus Christi, Texas and another night. The trip from Corpus to San Diego was long time-wise, but that was all right and the weather was holding.

I should have known that something would happen because everything was going so well. I had climbed the plane up to 10,000 feet to clear the upcoming mountains. it was beautiful—we were about 20 minutes southeast of Tucson, and listening to what we called back in New Jersey, Cowboy Music. All of a sudden, I heard what sounded like an explosion. The instruments were all checked, and were normal.

I called the engineer in the tower who also had engine instruments; the sound definitely came from the area of the left engine and just as the mechanic was telling me, he had no indication of any problems I could see heavy smoke coming from the port engine and the mechanic saw it at the same time. We determined that the engine was not on fire and what we were seeing was oil, which we figured was caused by the failure of a seal between the power and blower sections of the engine. I decided not to feather the engine and set it at about 1200 rpm, which would be a neutral position not causing any drag, and not providing any thrust: this later turned out to have been a very good decision. We had up around Phoenix two potential water landing areas one east of Phoenix: which was about 30 miles long, and one northwest of Phoenix. We immediately headed toward the large lake and opened the emergency packet that we all carried and to our dismay found out that the water in this lake was very low at this time of the year and not recommended for landing. We turned the old bird around and made for our only other choice, Pleasant Reservoir about 220 or so miles of Phoenix.

We were in for a surprise when we got there. This was an irrigation reservoir and it was hard to guess how much of the water was useable. The dam was made from a mountain that came down from the north and spilled into a reservoir. The dam was constructed across the Y area. The other or north end was very shallow and a large number of tree stumps could be seen disappearing into the water. I decided to make the approach over the dam to the north toward the shallow water.

While investigating the area we also noted a man fishing in a rowboat. This was a remote area. We had been talking to Phoenix radio all this time and asked them to keep the Naval Air Station at Litchfield Park, a few miles west of Phoenix advised. The lady at Phoenix Radio came back and said that Litchfield Park requested that we land at their base. I requested that she thank Litchfield for the offer and tell them I just don't have any wheels and would appreciate transportation from there at Carl Pleasant Reservoir. We all got a little chuckle out of that.

The time had come to land this crippled turkey. I made a standard counter-clockwise approach only because this was what I was used to and felt most comfortable with: downwind, base leg the final approach. On final approach, I knew I had to clear, the top of the dam just as close as I could possibly make it. As we neared the top of the dam there suddenly loomed a crane on its top. The crane rose up 35 feet above the arm and I was aiming for a spot much lower than that, so the decision not to feather the port engine became a life saver. I poured on the coal to both engines and the port engine answered the call. The port engine, however, with the power on it threw out a long black stream of oil. The guy in the boat was cutting a large white wake across the lake at this time. There was no wind so there were no ripples on the water so all one got was reflections from the top of the water. Our next and final attempt was made with Lt. Moore flying the course and keeping us heading straight down the longest straight line we could use and I handled the attitude mostly by watching the water line along the short. The landing was smooth and satisfying. We taxied to an area near shore and dropped our anchor, and pulled the toggle on our rubber there. The guy in the boat told us that he thought we were on fire on our first pass, and thought he would be a little more comfortable watching the next approach from the shore. We were eventually picked up and returned to Litchfield Park, NAS. I made it home about two days before my second child, a boy, was born.

# Women's Airforce Service Pilots (WASPS)

## Betty Pfister

Betty Pfister's introduction to flight came at a small air show. At the time, she was attending Bennington College, and she and a friend decided to go to the event. After a flight, she says, "I knew what I wanted to do the minute that I landed. There was no question." Immediately, she called her dad and said she wanted to leave college. The thought panicked her father who pleaded with her, "Please don't move, don't leave, I'll be up tomorrow." Her father came up from New York, and the two conferred and reached an agreement: If Betty would stay in college and keep up her good grades, her mother and father would pay for flight lessons.

After getting her private license, Betty was working on her commercial license when she got a telegram from Jackie Cochran. Cochran had gotten the name of every woman who had a private pilot's license or better. The telegram inquired if Betty would be interested in a training program. Betty finished college early and started with the WASPs in February of 1943.Betty Pfister was born in 1921 in Great Neck, New York, and lived for many years in Aspen, Colorado. She was a member of the 99s, the Women in Aviation International, the WASPs, the Colorado Hall of Fame and the Whirly-Girls.

At age 70, Betty served as a member of the U.S. Helicopter team. A woman of flair, she would buy a Bell 47-G and have it painted like a butterfly and called it "Tinkerbell." She received the Federation Aeronautique Internationale Rotorcraft Gold Metal for support of world-wide helicopter flight. Recipient of the Congressional Gold Medal for her work with WASPs, she was qualified in more than 25 airplanes. A philanthropist and support of young, aspiring fliers, she died November 17, 2011 at age 90.

*How did you get interested in flight?*

I had a brother a year older than I was, and he started flying. He went into the Naval aviation at the start of the war, and then was killed, flying off a carrier. I was in the WASPs at that time, in the middle of my basic training.

*What were your feelings? That's enough to make somebody quit.*

I almost did. I went home for a couple of weeks, and my family had a big conference. My brother was the oldest one and the only boy. I had one younger sister. Finally, I guess they just decided that it was the thing I did best and I loved so I contributed to the war effort. They let me go back and finish my training, and I have always been happy because it has been a source of great pleasure to me over the years.

*Tell me about your brother. What was he like? Tell me a little bit about him.*

Well, he went to Yale University and he was a terrific guy. We had a lot of fun together, and I don't know what else to tell you.

*He was coming off of a carrier? What happened?*

It was the catapult that failed to work and the Navy Torpedo bomber he was flying just went over the edge of the deck, and that was it.

*Where did you train?*

The first three classes of WASPs trained in Houston, Texas, but from then on, all of the training was done in Sweetwater, Texas. I was in the fourth class, and we were the first class to train at Sweetwater.

*Did you have any unusual things happen during training?*

Yes, I had just one thing. On my first day, I went up in a twin engine, UC-78, and there was another girl with me as co-pilot. I was to fly about an hour, and then trade places. About fifteen minutes after we first took off, we lost an engine, so we turned around immediately to go back to Sweetwater Field. About five to ten minutes later, we lost our second engine. So we, in effect, had become a glider. We just barely made it back to land. We did not land on the runway, but on a nice, flat dirt place. We probably totaled the airplane, but we weren't scratched at all. What had happened, they found out later, was that the man in charge of fueling that particular plane said the electric gas gauges showed that the tanks were nearly full. Our instructor had followed us and checked the plane out before turning us loose, and he felt we had plenty of fuel.

*What did you say? Did you have a talk with your instructor?*

No, no, no. He was just glad that we made it back to the airport and that we were all right. He couldn't have done anything differently.

*Describe the landing a little more specifically.*

There was a very steep embankment of rocks. I can practically see it in my mind. We had already lowered our gear, and we still had one engine going, and were getting ready for our landing when the second one quit. We wanted to go as slow as we could, and we needed every bit of altitude to keep us going above this big embankment. We finally just slowed up just enough to clear those rocks and sort of crashed on to the field, but from a very low altitude—but from maybe two or three feet. It did an incredible amount of damage to the aircraft.

*Did you have any trouble getting out of the plane?*

Not a bit, no.

*You just climbed out?*

Yes, but two WASPs had been killed the day before, so everyone was pretty edgy.

*So then what happened?*

Well, the ambulance came out. They had one on the field at all times. The poor flight surgeon had only been there about a week, and on his first day there, the fatal accident occurred. Then we called in with an emergency, ad so he was pretty upset and pretty glad that we were fine.

We all worked hard and we loved to fly and some of the ladies, the girls, washed out and that was sad for them and sad for everyone.

*Did you have any friends that washed out?*

Oh sure, there were some in my bay. There were six girls to a bay, and we had one bed, I think, that nobody wanted to sleep in. This was crazy, but we were a little superstitious. I think three girls in a row that slept in that bed washed out.

*After your training, where did you go?*

I went first to Wilmington, Delaware, to New Castle Air Force Base, and ferrying a lot of planes. From there, I transferred to Palm Springs, California, which is a wonderful base. I was there, totally, out of season—I think it was something like May through September and it was extremely hot. It was so hot that we had to do all our flying between five in the morning and ten and then quit until about five p.m., because the heat burnt up the engines and the brakes and everything. It was pretty nice. You would just go to the swimming pool and the Officer's Club every day, and from ten to five, we had time off. We got paid for sunbathing, which was lovely.

*So what did you do? You just developed that tan, right? Did you meet young men over there or not?*

Oh yes, there were a lot of guys there. I would say something like seven hundred men and thirty women, I don't know exactly, but you could have been cross-eyed and bow-legged and had a date every night.

*Did you date over there?*

Oh, sure.

*You didn't marry anyone out of there did you?*

No.

*How close were the women in the WASPS?*

Oh, very close. Physically close. Even during training we had six girls to a bay, six in a bay, and then we shared a bath, with two showers, two toilets, and so forth, and then six girls on the other side. So we were physically close and, I would say, close emotionally, too.

*How many of those friendships have sustained themselves over the years?*

Oh, a fair number. One of my best friends also goes to reunions, and I get to see her. We have some wonderful reunions every two years. We have conventions, and it really is fun. You look across the room and you see a face that even fifty years later looks familiar, and you can't possibly think of the name. Of course, a great many of us got married and that changes the name completely, too. Still, you can say, "Ah, you were with me when that B-25 caught fire that night over Chicago," whatever it was and see things you never will forget. I am just making up the bit about the B-25. As for my co-pilot, the one that was with me when I crashed the twin Cessna, I saw her about forty years later. I was sitting at a head table at a luncheon, and she recognized me. She didn't know my married name, but she recognized me and came up to me and said, "Didn't I fly with you that day when we crashed a twin Cessna?" and it was exactly the same girl.

*How many planes do you think you delivered?*

Oh, I couldn't even begin to tell you. I'd have to go into the logbooks and add them up. A lot. We did an awful lot when we first got to Wilmington, Delaware. In fact, we delivered a lot of primary trainers. We'd pick them up in Hagerstown, Maryland, and deliver them to all kinds of places. The planes had almost no instruments in them, very basic instruments, not much. The PT-19 was an open cockpit airplane, and was it cold in the winter! We had these big teddy bear suits, big sheepskin lined pants and jackets and even boots. They were very clumsy, but very helpful in the cold.

*You flew helicopters.*

Not with the WASPs, but afterwards. It was 40 years ago that I got involved with helicopters. I got my helicopter license in 1961.

*What number helicopter license are you?*

I'm Whirly-Girl #52.

*What interested you in helicopters?*

It was strictly by accident. I was going to get my ATP, my air transport rating. I had taken the written, which was a really tough one. I passed that, but it's only valid for six months. I wanted to take the flight part of it while the written part was still valid. I heard of a wonderful flight school in Fort Worth, Texas. I was in Fort Worth with my husband because I was going to start in flight school and finish my ATP. One of his customers was a man at Bell Helicopter, so we were out there one day. Normally, I would just sit in the car when I would go with my husband, which was very seldom. My husband said, "This is an awfully nice guy that I call on here. Why don't you come in with me and meet him" And that seemed to be a nicer thing to do than to just sit in the car and read a book, so I went in, and this man and I talked for a bit. He wanted to know what I was doing, and I explained that I was there to start training for my ATP at the flight school the next day. He didn't say anything else, but he picked up his phone to call the flight line. He told them that he had a lady pilot, but that she had never been in a helicopter. And to take me for a ride and let me try and fly it. And we did just that.

I was wearing high heels, I remember, and first thing I had to do was to get those shoes off. We went up, and I piloted for about five minutes. We probably flew for about 45 minutes. I was just terrible, which everyone is when he or she first tries to fly a helicopter, but I just fell in love with it. We made a bunch of landings out in the field where some horses were grazing, and you could see grasses blowing. Somehow the horses never missed a bite; they just kept munching away. Then we came back and landed, just about exactly where we started. I knew right then that I wanted to learn to fly helicopters, and so I did. And I went to Oklahoma and found a wonderful woman instructor. It was too expensive at the Bell plant. They had instruction there, but it was $75 an hour, which is about like $700 today.

*This was pioneering stuff, though, since helicopters had not been around all that long.*

The Bell 47, which was one that I think they were producing the most of in the Bell plant, was just a wonderful machine and a pleasure to fly. It is the same one as you see in "M.A.S.H." with that big round bubble. In Oklahoma City, we also flew a helicopter called the Brantley. In fact, I think I'll start a club called "The Brantley Survivors Club." It was an awful, little, two-blade helicopter. It was so squirrelly, and if you learned on that, boy, everything else was very easy. I didn't really have much good to say about it, except that it was easier to operate than the Bell 47.

*Did you have any close calls with the Brantley?*

No, not really.

*Did anybody else that was training with you have troubles with it?*

No, there was just one guy that was there at the same time that I was, and we both took our flight tests the same day, and both passed; however, the FAA gentlemen who were giving the flights tests saw Brantleys crack up right and left. It had a not-very-good reputation.

*What made it so difficult?*

The Brantley had to be landed absolutely. Forward motion was the only thing that was okay, but heaven forbid you did any sideways motion or any rear, because such motions would be just awful. Also, it had something else called ground resonance and though you could land with a little bit of forward motion, the best approach was just absolutely straight down. With anything else, it could self-destruct. It was not a very easy thing to fly.

I*n some ways it was a good start because, you knew that it couldn't get harder.*

Exactly. Everything else was relatively simple. I remember I smoked at the time, and I remember asking Dottie Young, my instructor, "Now where is the ashtray?" and she looked at me and said, "You are going to be a lot too busy to have a cigarette. If you want to smoke, you'll have to land somewhere." She was also a smoker.

*Tell me about Dottie Young.*

She was an outstanding instructor. I don't think I could have had a better one. She was about five feet tall, probably weighed a hundred pounds, if that, and was just a wonderful, wonderful pilot. My husband got his rating from her also, quite a bit later.

*Did she have any tips or any kinds of things that she really stressed that made a difference down the line?*

Oh, all I can remember is her telling me, "Now don't ever do that again," and I'd look at her and say, "I'll never do that again." She was very emphatic about what would and would not work for the Brantley.

As soon as I finished the flight test for helicopters, Jean Howard in Washington, the one that started the whole "Whirly-Girls" organization, called and said I was number 52. There were only that many women that we knew about. There were many more in Russia, but now I think we are up to eleven hundred and something.

*Did you have any close calls with a helicopter?*

Not really.

*Nothing, huh? Never?*

I had my own helicopter up until about fifteen years ago. I had a Bell 47, which I just had the most fun in. It was really a great running machine.

*A lot of people think of helicopters as capricious and...*

Dangerous.

*Is that a bum rap?*

No, they are fairly noisy and they have an awful lot of moving parts. They are quite a lot harder to fly, in my opinion, than an airplane. If you're not handling an airplane very well, as a student, the best thing you can do is take your hands and feet off the controls and it will come back to level flight by itself. But boy, in a helicopter, that isn't the same thing at all. It will crash in short order if you just leave it alone. Helicopters are hard to fly, and I think that makes it more fun. They are a real challenge, and the first day, you kind of get ahead of the helicopter instead of the other way around—it's very rewarding.

*How long did it take you to get to that point?*

I can't remember how many hours I had. When I got my rating, it was too few, but it was a legal amount, whatever it was, something like fifteen or twenty, which wasn't very much. It was about enough to get into trouble.

*Did you do commercial work then with your helicopter?*

I didn't do any commercial work. I had it just for my own pleasure really.

*Did people say, "How in the world do you that just for your own pleasure?"*

It was very expensive because there was no revenue coming in and they're expensive to operate. But it was all that fun.

*Tell me one of the fun trips that you took or one of the most memorable experiences you had while flying it.*

Well, we have a ranch at Woods Lake that is at ninety-four hundred feet. It's only accessible in the summertime. We get in there usually about the first of June and shut down about the end of September. Anyhow, it used to take about an hour and a half or two hours to go there by car on the roads that I find terrifying, so I was very happy to have the helicopter and make it in about fifteen minutes.

*So, you find the helicopter safer, you feel safer than riding on he road.*

Oh yes.

*What did your parents think of your flying? Particularly later on when you did the helicopter?*

They thought that was fine. I had a little fighter plane after the WASPs was deactivated, though I never got to pursuit school. In fact, I got measles the day before I was supposed to go. Twenty-one years old, and suddenly I broke out in measles. As a result, I spent a week or ten days in the infirmary, instead of going to Brownsville, Texas, to pursuit school. That was the one thing I wanted more than anything was pursuit school. The only problem was that that was the last class scheduled to go to that school, so I couldn't just go to the next one a month later. When that happened, I called everyone. I think I called General Arnold—everybody I could get a phone number for and just begged and pleaded to go to the next month's school. So I didn't get to fly fighters. I flew quite a few bombers, twin engines and four engines. After the WASPs were deactivated, a friend of mine and I bought five AT-6 advanced trainers that the military were through with. At that time, they were available for civilians to purchase. I think that we paid about fifteen hundred for each plane.

We would fix them up and get the civilian license on them and sell them for about three thousand. We made a small amount of money, maybe, four or five hundred dollars on each one. We kept one for ourselves—to fly. Finally, that one got too expensive to fly. I think it burned about thirty gallons an hour, something like that, so we finally sold it, and that meant that I didn't have anything to fly. Then I saw this P-39 Bell Airacobra sitting on the ramp at my home airport. I looked at it for a couple of days and then went in the office, and the secretary gave me the owner's name and phone number. The owner, it turned out, had bought it at the Cleveland Air Races and had flown it to New York, or Westchester County, which is where I saw it, and had scared himself to death. He had never flown anything as close to that powerful, so he had it for sale for a thousand dollars. It was a beautiful aircraft! The P-39 was a single-place pursuit plane, so the only check out you have is someone standing there, telling you what the different controls are and what to expect. And I will never forget, when I told him I was interested and wanted to buy it, but I didn't know for sure if I would like it. I had never flown one before. He said, "Well, why don't you give me your check ahead of time, in case you crack up."

*That makes you feel optimistic, doesn't it?*

Then he said, "I will give the check back to you if you don't like the plane and just charge you for the flight time." I think it is forty dollars an hour and that seemed very fair to me, so I went and got a cashier's check and handed it to him. And I had never flown anything like it. The P-39 was something like a thousand horsepower. There was quite a lot of difference between that and the AT-6, with 650 horsepower. Anyway, I took off, and I remember I forgot to pull up the landing gear. At the time, I was only going about two hundred miles an hour, and I expected at least two-fifty. Finally, I realized I had forgot to pull the gear up. I flew the plane for a half an hour and just loved it. It was the nicest little plane. Then I came in and landed, and told him to keep the check. For about four years, I flew in some air races and had a wonderful time. The P-39 and the Bell helicopter were the two best aircraft I ever owned.

*Tell me an air race story.*

I did have one thing. I was flying a Ryan Navion in what was called the International Air Race. It started in Canada and went down to West Palm Beach, Florida. Just before I took off from Montreal, a friend of mine gave me a big chocolate milkshake and it was in one of those cardboard cups or containers. I think it even had a lid on it. But anyhow, somehow it spilled on takeoff and that plane smelled so awful. That chocolate milkshake was all in the carpets and everywhere. I was flying solo and got in some really bad weather, and the only thing to do was find a place to land because there was no airport anywhere nearby. I finally saw a big racetrack. And so I landed in the infield. Just barely had enough room. The plane had a hand brake and I still remember pulling on that brake as hard as I could, beginning from the far end of the field. I had no idea where I was, and then I saw a dog by one of the little buildings. I went in, and I found that I was in Alfred Vanderbilt's private racetrack. There were a bunch of guys playing pool there. I had to call the FAA and let them know where I was and that I was fine. Here I was, smelling of that milkshake. I can still smell it today. Finally, the ceiling lifted a little bit, and so I flew on down to either Fort Lauderdale or Miami, to finish that race. And I won the race. Apparently, I had been on the ground for a shorter time than anyone else.

I raced the P-39 a few times, and that was always kind of memorable, just because it was the only one. But they didn't allow women to race at the Cleveland Air Races. They were the big deal in those days. So I had to have a man fly my plane. We didn't do very well because there were a lot of P-51's, and they are just basically a faster aircraft.

*You must've been pretty frustrated that you couldn't fly at the Cleveland Air Races.*

Well, I was.

*How frustrated were you?*

Well, I'd love to have flown it myself, but I found a pilot that liked the P-39. Not everyone liked them. Chuck Yeager particularly liked it, I remember.

*Tell me what Chuck Yeager did.*

I was flying in an air show there at Cleveland. They were paying me in a three-day air show to fly the P-39, supposedly doing a mock girl fight with a friend of mine who was flying an AT-6, which wasn't nearly as fast. It was a foregone conclusion that I was going to win this little dogfight. During our act, I didn't do anything too spectacular because I wasn't that good a pilot. I was okay, but nothing like Chuck Yeager. Anyway, Chuck saw my plane and asked me if he could fly it in my place. I think it was the third and last day at the air show. We were going to get him a wig so anyone who saw the plane taxi past would think it was me, but we never quite got that act together.

*That's pretty funny.*

We made an arrangement that Gloria, the girl flying the AT-6, would do all the radio for a flight of two because we didn't want the tower to hear Chuck's voice on the radio. So that was easy. And I will tell you, on take off, I just sat on the grass and watched. I think he did three snap rolls on take off and it went on from there. It was so spectacular that he announcer went crazy. He said, "Look at Betty flying! Wow! I've never seen a woman fly like that!" and he went on and on and that was fine for about ten or fifteen minutes, whatever the act was. Finally, when Chuck came in and landed, I suddenly thought, "Oh this isn't so good. They may be mad at us. They were paying me to fly, even though he did much, much better than I ever could have." I ran down to the plane as he was just getting out, and I said, "You know, Chuck, this isn't so funny now. We thought it was. But I wonder if somebody will get really mad." And here came this big red convertible, with the top down and all these generals and people in it, and they were becoming down to congratulate me on what a wonderful flight. And so, Chuck got out of the way, and I looked as though I had been flying for twenty minutes cause I was all hot and sweaty from running down the ramp. Chuck got the parachute halfway on me, and then the car arrived. And they kept bragging to me about how beautifully I had flown and I didn't say a word. Then they put me in the car—we were a long way from the grandstand—and drove me back there, but Poor Chuck had to walk.

*That's priceless. Have you been in touch with Chuck in more recent years?*

Yes, I see him from time to time at different conferences and things.

*Does he like to recount that story?*

He'll never forget it. It was really funny. He had good things to say about the P-39, where a lot of people don't. It had a reputation, which I never thought it lived up to, for something called tumbling, which is the opposite of rolling, and it's nose over tail, tail over nose, whatever you want to call it. I think that if you just get it too slow, you might have problems. For example, a lot of the Russians, who never had any training at all, would do that. Though I never delivered the P-39 to the Russians, because I wasn't flying pursuit, a lot of WASPs would deliver the P-39 to the Russians. I think Great Falls and Anchorage were the places where they landed them. The Russians would have some major problems with the tumbling, which is pretty hard to get out of.

*Why do you think you were so interested in flying a pursuit?*

I just kind of loved the idea of a solo experience. You don't have to get up in a bomber with a whole big crew, two, three, four people. I am kind of a fighter mentality, instead of a bomber mentality. But I enjoyed flying some of the bombers I checked out in. These included the B-25 and B-26. Also, I actually checked out in a B-24 and was co-pilot in a B-17 quite a bit. And it was fun. I didn't like the B-24. It was extremely heavy on the controls. We had two girls flying it, and then we would have a flight engineer, sergeants usually, some old guy, maybe thirty, thirty-five years old, married, usually with kids. Boy, these guys didn't want to fly with two girls at all! Still, after awhile, when they saw that we had a better accident record than the men, they were very happy to fly with us.

*What's the worst treatment you've received from a man?*

I was really lucky, I guess. The worst was when I left Palm Springs and went to Williams Field in Arizona. I spent the last three months in the WASPs there, but our commanding officer there, Col. Grill, did not like women pilots. There were only about ten of us there. It was a small group, but our job there was interesting. There were a lot of men cadets—I don't know how many—maybe four hundred, and they were going through their normal training, and a lot of them cracked up planes. The school was a class for advanced training. When a plane was cracked up, then it was supposedly repaired, and one of our ten WASPs had to test fly it and find out if it was okay for the men to fly. Everything was wrong with those aircraft. Most of our good mechanics had gone overseas, and boy, if your gear didn't come up or come down, you weren't surprised. That was almost normal. If the wings stayed on, you were lucky. And if the engine didn't quit, I mean, it was almost that bad. I think 90 percent of our test planes needed more work.

*Oh, that's tough, isn't it?*

Well, I didn't care for it much. You had to really stay on your toes because you didn't know what was going to happen next.

*What sort of things did the colonel say? Disparaging remarks?*

He did not say anything in particular. At Palm Springs, I think we had twenty-five or so WASPs, and the commanding officer thought we were great. We did the job as well or somewhat better than the men and they treated us beautifully there. And then suddenly we got sent to Williams Field. It was a whole different atmosphere. We didn't feel a bit welcome. That's where we were when we heard we were going to be deactivated. Everybody was deactivated the same day, December 20, 1944.

*Yeah, tell me your feelings about deactivation.*

We all just hated it. We would have flown for free, let alone for the very meager pay, looking back, that we were getting. I don't even know what it was. But we loved flying the military aircraft, and then suddenly to be told, "Hey, it's over." It was very difficult.

*Talk about your transition from flyer back to civilian society.*

I was lucky in some ways because I got a few jobs. There were a few flying jobs, not very many. But I flew for a couple of non-scheduled airlines. I flew from New York down to Puerto Rico. I flew co-pilot on DC-3's mostly. Then we did some trips flying Brahman bulls down to down to Cartagena, Columbia. We took nine at a time—three rows of three each. That was really not quite legal because there was a law that said, when flying livestock, that they should all be able to lie down at the same time. If you are know anything about cattle, you know they sort of get fatter when they lie down. So on our flights only two could lie down and the other one would kick the heck out of the ones lying down. And when it was his turn, he would get kicked. At the time, the company was scarcely making any money as it was. With eight cattle, we would never have been able to get to fly the flight. We flew out of New Jersey and we would go to Lumberton, North Carolina, because fuel there was something like five cents a gallon cheaper there. For that reason, all the "non-scheduled" flights used Lumberton, and we would go to Miami and then on down to South America or Central America.

On this one trip, I felt really sorry for these Brahman bulls. They landed in Miami, and it was very hot and the bulls had been cooped up for a long time. I was just the co-pilot, but I asked the pilot if I could take them off the aircraft and give them some water to drink before we took off. And he said fine. It wasn't so easy to get them off. They all were easy to lead on board, with a halter on, but getting them to turn around, was difficult. Then they practically galloped. You had to get out of the way when they saw that fresh air coming in the back door. Outside, I tied maybe one or two at a time to the metal fence. Then I would get buckets of water, and they were really happy to get a drink.

But at one point, I didn't make a very good knot, and the next thing I knew, the bull was loose and just galloping down the runway and all over the airport. There was a great deal of traffic in Miami Airport. They had Eastern and American Airlines there—you name it. And so the tower had to shut the field down for, I think it was, about two hours. The airport authorities finally got a professional cowboy in a truck with his horse, and he managed to lasso the bull and we loaded him up again. But it was pretty exciting for a while.

*Did you get in any trouble for that?*

I didn't get in any trouble, no. My pilot thought it wasn't the greatest thing that I hadn't tied a good knot, but, anyway, it worked out.

*Now what's the name of this non-scheduled airline?*

I flew for one called AXITO. I couldn't tell you what it stood for. And then we'd go bankrupt, and the next morning, with the same pilot and the same aircraft, we would have a different name painted on the side. One was Puerto Rican Air Transport. We flew a lot of Puerto Ricans from San Juan up to Harlem. They would have some relatives in New York. In Puerto Rico, they had a lot of farms and they had fresh air and sunshine for their kids and chickens and cows. Then they thought America, or I should say the New York area, was the promised land. They'd come up in January, and the women were wearing sleeveless cotton dresses and it was just terrible. At one point, we were trying to get more passengers. I know we were giving this man a free overcoat and Eastern Airlines was charging something like a hundred dollars. We were charging something like eighty or ninety. We kind of had a price cut so we could get some passengers. We'd have a mother, we'd have a four-year old on her lap and a two-year old and then maybe a baby, and the one rule we did obey was seatbelts for everyone. That would take another half hour before take off just to get everyone belted in.

*So you were, in a sense, an early commercial pilot, weren't you?*

It was non-scheduled, but you could say that. The pilot I think was getting five dollars an hour, and I would get three dollars an hour.

*What year was this?*

Well, right after the WASPs, and that broke up in '44, so about '45 or '46, I guess.

*And you didn't have any adverse or unusual reactions to you as a female flying.*

No, I was a good pilot, I think.

*And just did the same job that a man would. But there wasn't any prejudice back then that you saw?*

Not for the jobs that I had or the area that I was flying in. I remember when the WASPs broke up that I wrote about fifty letters to every aviation recruiter of the airlines, but they weren't hiring any women. Of course, now there are a lot of women flying for the airlines, but in those days, there weren't any. And then I wrote to a lot of corporations that had aircraft, to see if they would like to hire me to pilot, but I never knew there were so many ways to say no.

*What was the lamest response you had from these people?*

Oh, I don't know. All the responses read something like, "Well, we are just not hiring anybody at this time, but contact us again in another six months or a year." Those were the nice ones, and that was it.

*Well, have you always liked a challenge?*

I guess so.

*Where do you get that? Do you get that from one of your parents?*

No, not really. I was really glad my parents let me continue flying after my brother was killed. I've enjoyed it so much all my life, and I'm still interested in flying. I'm going to a conference called Women in Aviation next month in Reno, Nevada, and I'm going to be honored there in something called The Hall of Fame for Pioneer Women in Aviation. And all of us are getting on in years as you can realize at this point.

*Can you imagine yourself not having flown?*

Probably not. I have no idea what I would have gotten involved with. Well, I will be eighty years old in about four months. And they had a little air circus, but I thought it was just marvelous. I knew what I wanted to do that minute that I landed, there was no question, it was just so black and white. And I called my dad in New York and said "Hey, I just went for an airplane ride and I want to learn to fly, and I want to leave college and start flying tomorrow." And the only problem of course was money. I didn't have any money. I kept thinking my parents would probably have given me some sort of very minor allowance, for whatever I was spending it on, but nothing like an actual flight lesson, so I kind of panicked my dad, I guess. I was doing well in college and he said, "Please don't move, don't leave, I'll be up tomorrow." It's a long four or five hours from New York, and he came up and we had a big talk and came to an agreement. If I would stay in college and get good grades, he and my mom would pay for flight lessons. And I thought that was a wonderful thing to do. And that's what I did and that's how I got started. And then when I had about 150 hours, and I was working on my commercial, which takes about 200 hours, I got a telegram from Jackie Cochran. I had my private license, but not the commercial, and she had gotten ahold of every woman that had a private license or better and wanted to know if we would be interested in a training program, very similar to what the cadets, the men cadets had. I had my return telegram back before you could bat an eyelash. I think a great many of the other girls did too.

*And what year was this?*

I started in the WASPS in February of '43, that was my senior year of college and I finished college early and wrote my thesis early cause they were letting us do all kinds of things because of the war. And I started in February of '43, and we were all disbanded in December '44. So I had almost two years, which was great.

*So how many hours do you have total?*

Something around nine thousand now. My husband and I did a lot of flying and my license is no longer current. I quit flying, I'm trying to tell you when I sold the helicopter, that was the last of my flying, and I think this was between ten and fifteen years ago. My memory is not great for some of these dates. They're not going to be accurate.

*Well, that's amazing though. Wonderful. How many air vehicles have you flown and checked out on, do you suppose?*

Counting different models of the same plane at the base, we had the BT-13 and the BT-15, they were very, very similar. One would have a slightly different tail or something. But they were practically the same thing. I guess maybe twenty or twenty-five.

*Wow. Do you have a favorite Jackie Cochran story?*

Not really.

*Did you really like Jackie very much?*

Not at all. But I was always very very grateful to her and I am today because without her we never would have had the WASP program.

I still remember the night my brother was killed. My mother called the base, I was in the middle of basic training at Sweetwater, and we were allowed to go outside of our bay, our room that we all lived in, I think until ten o'clock at night, or something like that, and then we had to be in the bay, if not in bed, I've forgotten the rules. But I went for a walk and that was perfectly legal. And then when I came back, one of the bay mates of mine, said "Oh Betty, you had a phone call and it was important." But they couldn't find me, so they were going to talk to the commanding officer. Well, that was looking so bad, I just knew it. And the baddest thing I could think of was that something had happened to my brother. Otherwise, why would the commanding officer take my call? And Jackie Cochran was on the field at that point and she was talking, they were in a meeting, she and the commanding officer. And I guess some other people, and so I went up to the main room, which was quite a ways from our bay. And, that's where the telephones were and I was going to call my family and make sure, if anything, was okay. I didn't think it was, but I was hoping I was mistaken. And so, the guy that was in charge, the officer of the day, or whatever he was, read my name tag and he said, "Oh, you're not to make any phone calls." I thought "wow" and then I really knew it was something bad. And they went on with that meeting for maybe another hour. It was after eleven o'clock at night and, finally, Ms. Cochran came out and she and the commanding officer, I don't know which one of the two told me what had happened and that they wanted to be with me when I got the terrible news. And I always thought that was not great of Jackie, that it wasn't very understanding. You know, if you had the word, it would be, the sooner you know what's happened, the better. It was probably worst hour of my life.

*Oh boy. Yeah, that's incredible. Just incredible.*

# The Charter Life

## Holly Robinson

Holly Robinson was born in 1909, and at the age of five discovered flight when a bamboo and linen plane flew over Paragould, Arkansas, where he and his family were living—an event that he recounts in the interview. From his father, a railroad man he describes as "ornery," he gained a love of the outdoors. During the Depression, Robinson worked for Western Union as a telegraph operator and sometimes branch office night manager. His first flight training was in Santa Barbara, and his flight test for his first flight certificate with the C.A.A. (now the F.A.A.) was at the dirt field now called L.A. International.

After several years of flying missionaries and supplies into Mexico, he started a business, but in a few years, the dream of flying, instilled in his memory from his youth, won out. He sold that successful business and went to work for McKenzie Flying Service in Eugene, Oregon, as a charter pilot and instructor. He later went on his own, and when we interviewed, he was eight-six, still instructing and living the beloved life of flight.

An easy-going man, he was still enthused by life. Recently, while spending time at the Oregon Air and Space Museum in Eugene, he was surprised when one of the ladies, a mother of grown children, stepped up to him out of the assembled group and, in Holly's words, "put her arms around me and said, 'I love Holly.' I was so amazed. It just amazed the heck out of me."

*Do people ever look at you, surprised at how old you are?*

Yes, they do. The funniest thing I ever had anybody say to me was after I told them my face was completely reworked by a plastic surgeon in Mexico. One of my friends said, "Holly, you mean you weren't always that ugly." (Laughs)

*Why did you have your face redone?*

Because I had my face smashed in an airplane crash, in which I was not the pilot. If you use that, be sure and emphasize I was not the pilot. The crash happened in Mexico City. It couldn't have happened in a finer place actually, because I had the finest care this side of heaven there.

I had flown numerous flights, flying missionaries into Mexico. Have you ever heard the line, "The most dangerous thing in the air is a doctor in a Bonanza." I was with a doctor on this flight. Previously I had flown his plane without him a number of times. This time we had gone from Hermosillo to Navojoa, Mazatlan, Guadalajara, then to Mexico City where the disaster occurred. This was over thirty-five years ago. I had only about 400 hours as a private pilot at the time. What I didn't realize was that we'd been flying along okay at over 12,000 feet. He was barely on the job. What he actually did in his state was a high-speed stall, turning finally to the runway. We hit the ground at a forty-five-degree bank and at about 100 mph and only survived by the grace of God. As a consequence, we didn't stay two days in Mexico City—we stayed two months.

I wound up right underneath the wreckage, under the right wing. Our wives were sitting right behind us when this happened. That landing gear came up through the plane and nearly took my wife's left leg off—between the knee and the ankle. I was out of it; I didn't know what was going on at all. The ambulance picked us up and took us to this little butcher shop of a hospital.

There they were going to take my wife's leg off. Then the American Embassy heard about these gringos. They came and got us and put us in the best hospital in Mexico, the American British Cowdray. They even put my wife and me in the same room, which they wouldn't do in the U.S. We had nurses around the clock—twenty-four hours per day. It was a miracle that we survived. My wife and I fell in love with the nurse, and we brought her home with us as our daughter. As a result of this, we acquired three other girls, too.

*Talk about other flights in Mexico.*

After I'd fly down, I'd often take a missionary as well as often times a plane full of supplies. This one time I had the plane completely filled with shoes. In fact I was flying for the Seventh Day Adventists, who had a missionary school in Navojoa, so I stopped in Mexico to have the official examine the airplane. He looked in there and saw these shoes. He said, "Oh, this pair will fit my little girl." I said, "They're yours." He signed my papers and away we went.

*How'd you get started?*

When I was five years old and living in Paragould, Arkansas, I saw my first airplane. The airplane flew over and threw the town into an uproar. People were running out in the street all over town, shouting. "An airplane, an airplane." I was thunderstruck by the idea that people could actually fly, and I guess I never quite recovered. I still think it's the greatest.

It was a good many years later when I lived in Santa Barbara that I actually learned to fly. I learned to fly in a little two-cylinder plane, an Aeronca. My instructor was the late Harry Crosby. Harry died some years later flying an experimental plane for Northrup. By the way, this was over sixty-one years ago, because I've been married for sixty-one years, and I was flying before I was married.

I was going to show off—to show what a hotshot pilot I was. I was going to go up and do a loop. I had had seven hours of flying time at this point. When I was up at about 1,500 feet, this crankshaft broke in the front of the engine, and the propeller and the hub and everything flew out from the plane by itself. Everything forward of the crankcase went ahead of me, then it shot straight up and I passed under it. I looked back and watched it come down behind me. I actually managed to go back and land at the airport. I made a downwind landing, because when I left the airport there were two biplanes sitting there, which I though could cause me problems. When I looked back, I had noticed they were gone. I had enough altitude to fly a normal traffic pattern and land into the wind. But now I'm flying a glider. If someone cut me off in the pattern, there was no way to go around. So I made a few fast turns to lose altitude and landed downwind.

My instructor, Harry Crosby, came running out to ball me out for making a downwind landing. When I got close enough, I said, "Harry, did you see where the propeller went?" He stopped dead in his tracks and said, "Propeller?' That was a sad day for him, because that plane represented his sole way of making a living.

*Talk about Harry Crosby.*

Harry Crosby told me a story about when he ran a flying school in San Bernardino in the late twenties or early thirties. One day a big fancy car pulled up and a real pretty lady came into the office. She lay her coat on a chair and announced she wanted to learn to fly. Harry talked to her for a while, then went out and they got in this Eagle Rock biplane with a Hisso V-8 engine. Anyway, he finally got this young lady in the front seat of this plane, and they took off. At about 50 feet in the air, everything forward of the firewall disappeared. Of course, the airplane with all that forward weight gone, hit the ground tail first. Harry was kind of shaken up. He scrambled around and got out of the wreckage and looked for this girl, and she was gone. And he said, "Oh, my god, what happened to her?" Then he looked up, and she was running like a professional sprinter toward his office. He couldn't catch her. By the time he'd gotten there, she had grabbed her coat, got in the car, and disappeared. And he said he hadn't seen her since. He still had her $350 though.

What happened was that Harry's plane originally had one of the smaller Hisso engines in it, but somebody had changed it and put in a larger one. Because something didn't fit too well, the person putting it in filed out some holes in the engine mount. The accident occurred when the engine just broke out of the holes, and the whole engine and everything just fell out.

*You worked as a toolmaker for the aircraft industry in World War II.*

I worked for Interstate for a while. We were building a torpedo drone in downtown L.A. Then after that, I went to work for North American at Los Angeles International Airport. I worked on tooling the B-25.

*What were you thinking when you were working there? You were sort of involved with planes, but you weren't flying them.*

No, I wasn't. I didn't have any college degree, so out of high school I had to go to work. I did have a private pilot's license at the time. I tried to get in the Air Force. When I went to take my exams, I told my boss at North American, "I'm taking Wednesday off." That's all I told him. But he said, "I'll tell you what you're going to do, you're going over to Ontario to take the exams, at the airport." I said, "No, I'm moving. I want to be home to help my wife straighten the house." He said, "You're not foolin' me! I know what you're doing."

*How'd you feel?*

I kind of felt bitter about not getting into the Air Force. I know I wouldn't have been a fighter pilot. I probably would have been ferrying bombers across the ocean, or towing targets, but I would have been flying. That's what I like to do.

*Was it your age?*

No. I was at the very top age where they accepted somebody—thirty-seven, but that wasn't it. The government just wanted me to stay there and build the airplanes. Even though I passed the exams. As I said, I was kind of bitter for a while, but some of my friends that had gone into the Air Force, said, "Holly, you shouldn't feel bad about it. Somebody has to build these airplanes."

*Ever meet any bigwigs at the plant?*

No, I never met anybody at the plant, but years later, I met my private personal hero, Jimmy Doolittle. That was when I still lived in southern California. I had always just adored that man. At the time, I was on the aviation committee for the chamber of commerce at Inglewood. Doolittle was there at one of the meetings. When the meeting was over, I walked up and I said, "I'd like to shake hands with my personal hero." He thought I was going to say something about the Tokyo raids. I said. "I've been following you actively before you flew that little Laird racer." He looked kind of sober until then. Then he said, "Oh, you know about that little racer. I had a wonderful time flying that thing." About that time, somebody more important showed up and ended my little meeting.

*Tell me about quitting a successful business to go into flying.*

I was always crazy about flying. But I had started this business with this other fellow. We invented a little machine that would set the teeth in a handsaw. You know how the teeth are staggered, so they'll cut a wider path. The company was successful, but the year after my partner tried to kick me out. I ended up buying him out. Wife Mildred and I had bought property in Gardenia and had it for nineteen years until a fellow from Eugene, Oregon, Leo Nielson, bought me out. I came up to help run the business for one year for him, which I did from January 1st to the end of the year 1966.

When my year was up, I realized I had about 3,500 hours in as a private pilot. So I went out and got commercial, instrument training, and flight instructor licenses. Then I went to work for McKenzie Flying Service. I flew 800 hours in just one year for them.

*What did you do?*

I did a lot of fire patrol flying. I made many a turn with a wing ten foot off the top of the trees, but in twelve years, I never did have any forced landing up in the mountains. If I did, I'm sure I could outlive it.

*You're sure you could outlive it.*

I think so because my original flight instructor, Harry Crosby, had the ability to instill the proper kind of confidence, not overconfidence but the proper kind. And as he said, "If you always fly the airplane, you'll never get hurt. If you have to land in the trees, pick the smallest one."

*What's your proudest moment in flight?*

When I landed that airplane, minus the propeller, I think.

*What did people say about that?*

They congratulated me for not bending the airplane. Like Harry Crosby said, "You're not through flying till the hangar doors close." By the way, Harry was, in this order, my flight instructor, my friend, and my hero. In fact, he was a fabulous pilot. In an air show in Santa Barbara, he put on a little show with that little Aeronca. In a steep turn, he drew a circle on the ground with a wing tip.

*You were there?*

Yes, I saw it. In fact, he was wearing an old suit that he'd dyed blue that was too small for me. It fit him like lacquer. (Laughs) He not only did this circle with the one wing, but he did a crossover and ascribed a figure eight in the turf with the other. When he got back down, Harry Bowen grabbed a hold of him, took him behind the hangar, and he said, "You crazy son-of-a-bitch. Don't you care if you kill yourself out in front of all these people?" And I'll never forget Harry's answer. He looked at Bowen and said. "You got the wrong idea. You could grind that wing off up to the first hinge before you get in trouble."

*What was the field like?*

It was dirt and grass. As a matter of fact, the day after that for the next couple of days, I helped Harry finish recovering the wing tips and straightening out the wing tip bows. He told me, "I don't do that for free. I only do those things when I'm getting paid."

*Let's talk about your flight students.*

I mentioned the one that Paul Harvey mentioned on the radio. I had this one guy who, after about an hour of instruction, thought he knew everything. And he just got really smart-assed about it. We were at about 3,500 feet high, so I rolled the plane upside down, crossed my arms across my chest, and said, "It's all yours." And he panicked. When I got him back on the ground, he disappeared. I haven't seen him since, and it's been about twenty-five years.

When this guy first came out, he said, "Are we going to stay around the airport and make landings?" I said, "Why no." He gave me this little disgusted grunt. "Well, why not?" he said. I said, "You probably can't fly this thing within twenty degrees of a straight line anyway." Very few people can when they first fly.

I had this other student that was interesting. One day I walked into the office there and saw this young fellow standing there. He looked as if he had his hands on his hips. One of the pilots, Verne Faulkner, said to this young fellow, "Here comes the fellow I was telling you about," referring to me. Vern says, "Holly, meet Mike McKern." So I stuck out my right hand, and he put out his right arm. The arm had one of those hooks on it. Then I stuck out my left arm and saw that he had one on his left arm, as well. Both arms were off between the elbow and wrist. Anyway, I taught him to fly. What I didn't know at the time, until I saw it in the newspaper later, was that everybody had turned him down for lessons. I thought nothing of it.

One of my young friends, Dan Knight, had a Bonanza. I looked on the schedule, and all the training planes were gone. I said, "Golly, I can't let you go without getting in an airplane." And this young friend of mine said, "Here Holly, here are the keys to the Bonanza." So I went out and took this young fellow up with the hooks. I put him in the left sear, and we flew around for thirty-five to forty minutes. During the period of his lessons, I made him special equipment—a little harness with a little metal plate in front of his mouth to put the mike on. He'd then learn forward, press the mike, and talk on the radio. I also made him a little ring on the throttle. I soloed that boy in nine and one-half hours.

Then I taught a young lady named Donna Bell to fly. She had no left arm whatsoever. I soloed her in about twelve hours. Later, my wife and I went to her wedding. We didn't hear from her for about two years. She moved to Idaho. She had a daughter two weeks ago, called Mildred up, and said, "I named her Holly in honor of your husband."

*Any situations where people have responded to you because of your age?*

When I was flying for McKenzie Flying Service, three fellows from MGM Studios in Hollywood came in. They wanted a pilot to fly them down the coast to Coos Bay, and then up along the coast from there at a slow rate. They were looking for movie sites. Of course, I was sitting there —one guy on the right; two guys in the back—not saying anything. I'm not trying to ask questions. These guys are busy observing. Just before we got back to Eugene and got ready to land, the guy in front looks over at me and said, "How old are you, Holly?" I said, "Sixty-nine." He said, "Damn, if I'd have known I wouldn't have gotten in the same airplane." But the next year when they came back up, they asked for me again. (Laughs)

I have access to a doctor friend's airplane. One day I flew it over to Independence, where a friend of mine has a little airport, and his wife has the café there. I had a bowl of chowder and left the waitress a big tip, got in my airplane, and came home. Then I realized, what I'd done. He made a few comments about me, as an old bastard, but promptly forgave me.

*How did you feel in you so-called second life, the flight career? You had a successful career. Any noticeable difference in your attitude?*

I think so. I know I was delighted to be doing for money what I liked to do for fun.

*Can you envision a time that you couldn't fly?*

No, I can't. Oh, I know I'm not going to last forever—I'm smart enough to know that. But I've often said "I'm gonna fly as long as I live, or as long as I can, whichever the Lord allows."

# A Final Messaging:
## The Philippine Islands

## Joseph Brodecky

I met Joseph Brodecky by happy happenstance at a WWII exhibit at the Winter Park History Museum in Winter Park, Florida, and later set up an interview at his high-rise apartment. He was a generous, affable man, easy-going and serene. A former athlete, he was stooped some but seemed in remarkably good condition for a man of 91. Before the interview, we strolled about the outside grounds, where he shared some stories. Then we sat down in the high, rarefied air of his apartment and began the formal interview.

Brodecky served as a Marine Corps pilot in the Pacific Theater. After a period in the Philippines, he was scheduled to go to Okinawa to search and destroy kamikaze fighters, but the atomic bomb was dropped, ending the war. A graduate of Spartan College of Engineering in Tulsa, Oklahoma, he worked in the airship business, to develop new blimps at Goodyear Aerospace (Lockheed-Martin). He was there for 39 years, and his endeavors included the position of coordinator of chief design engineer systems. His training in multi-engine had prepared him for a career as an airline pilot, but paths divert and, as he speaks, he seems pleased with the professional direction life somehow took. Joseph V. Brodecky, this consummate gentleman, lived his later years in Winter Park, Florida. He died at the age of 92

.

*You were a replacement pilot.*

Yes, I was. I went into the Navy flight program in 1943. At that time, the Marine Corps was established, and I transferred into the Marine Corps. After I went to my operational flight training, I was sent to San Diego and caught a convoy to the Hawaiian Islands and then on an escort crew to the Philippine Islands.

*Where were you stationed there?*

I was stationed at the southernmost part of the islands, at Zamboanga.

*What were your duties?*

I flew support group there when we were clearing up the Japanese. First, they modified our planes. They replaced our nose with a series of 50 caliber machine guns that were aligned at a certain angle. That way, as we would make our runs, we would fishtail with all the guns going at the same time and would just mop up the enemy as we were flying over.

A lot of the time it was difficult to see the location of the Japanese, so our ground troops would send up flash shells to help pinpoint the area we were to hit. We flew low-level over the treetops and we just annihilated the whole enemy complex. That was our main mission.

We had a lot of other missions we classified as sorties. We also flew photographic missions. Another time we dropped surrender leaflets.

***Author Irv Broughton with Joseph***

*How did you do that?*

The leaflets came in bundles, and we dropped them through the bomb bay. There were a lot of outposts still hidden in the Philippine chain. The Navy had control of the seas, so our mission was to drop the leaflets and hope the people understood them.

*Was it scary?*

No, we were very used to it. On some missions we were flying over treetops. The Japanese would set up land mines along the way and set them off when we came in low.

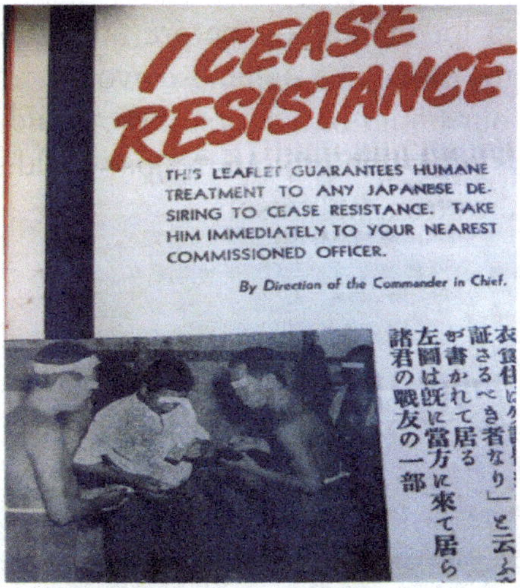


*Did you lose any planes?*

We had one plane knocked down.

*Did you lose any friends in that plane?*

Yes, we lost a few friends.

*Closest call during the war?*

In one of those runs, shells came up, and it looked as if we were flying through a rainstorm. A couple of the shells hit the right wingtip and damaged our radar scope on that side. That was the closest call I had. This was scary, but we made it home.

*Did you pray a lot?*

Yes, we did a lot of praying before and a lot of thankful praying after our runs. On the runs, the Japanese would hold their fire, but the moment we opened our bomb bay doors, they would let loose and shoot at us. Fortunately, that's when the ground troops would spot their location and come up and attack the stronghold itself.

*Any other close calls?*

It was in Livermore, California, in flight training. We had a series of lights to guide us. One time I was on approach from a solo flight and saw the lights, but when I came into the flight pattern, I noticed that a couple of planes were headed in from one direction and that I was coming in from another. Then I realized my mistake. I pulled up and got back into a pattern and landed. My instructor was there, and I got a very stern lecture.

*Were you, by nature, a cautious guy?*

I think I was conservative. I was raised that way on a farm. My father was Czech, and the family came to the United States from Prague. He graduated from the University of Nebraska and was a member of the Nebraska legislature. He raised strawberries in a town called Howells, which had a population of 952.

# War in the South Pacific

## Syd Trueman

Syd Trueman was one of not too many navel fliers that made two tours of duty in the South Pacific, which he spent in places like Guadalcanal, Vella Lavella, and Bougainville. A native of Birmingham, Alabama, Trueman was finishing his second year at Birmingham Southern when war broke out. He joined the military on February 16, 1942 and by March, he found himself in training in what turned out to be a "fast track." In January 1943, he was commissioned and went to further training, including a carrier qualification of eight landings on the *USS Wolverine* in Lake Michigan. The *Wolverine* was a side-wheeler with auxiliary power. He then headed overseas in July of 1943 and stayed in about eight months after the war ended. He married his childhood sweetheart soon after leaving the Navy. He was in the reserves until 1957, during which time he took up a career in sales, and then worked as an insurance adjuster for State Farm Insurance Companies. He retired in August of 1975.

*Why did you make two tours of duty? Were you a strong patriot?*

I was very patriotic the first tour—not so eager the second tour. I was told I had to go on the second tour since I was single.

*You met Eddie Rickenbacker once, I understand.*

Yes, I did. This was early in our training. He came up to the Atlanta E-Base and gave a lecture. He told us that a lot of things had changed, but one thing hadn't. He said that we "had to keep our head on a swivel at all times." I listened to him carefully and was thankful more than once for doing so.

*You don't have a swivel neck now, do you?*

No, it's not nearly as "swivel" as it used to be. It's gotten kind of creaky.

*Anything crazy happen during training?*

At one time, I did a count of fellow pilots I saw killed in my travels in the navy. I quit counting at thirty-two. I saw thirty-two guys die! Early in training at Jacksonville, I was playing touch football with some mates in a big grass field adjacent to the runway. A Beechcraft two-engine was landing when something went wrong. The pilot swerved and came down our way into a ditch and exploded into flames. Before we could get over there, the pilot and copilot were incinerated. It was horrible! Several of us were walking along one day at a field in the South Pacific when we saw a midair collision. We looked up and saw the F4U Corsairs collide. One plane disintegrated. One pilot got out and parachuted to within about fifty feet of the flames, and survived

Hellcat

*You fought in the Battle of Rabaul.*

The Marines were up high in F4U Corsairs and we were in Hellcats. As we approached Rabaul, we dropped our belly tanks. The left us a full load of fuel, but that plane burned a tank every twelve to fifteen minutes at combat power. We went up to about twenty-two thousand feet. Some Japanese already were engaging our guys below, and some were climbing to meet us. I saw what I though was about 230 Japanese planes. I was scared. As the enemy climbed, one of our men said, "Request permission to attack." Our skipper replied, "Stay in formation." Soon the second request came asking for permission to attack. Same reply from the Skipper. The third time he was refused, the fellow said, "Jesus, Skipper, what do we got to do, look 'em in the eye to shoot 'em?"

We finally got the okay to break off and go after targets of opportunity. I spotted a Japanese Zero below me and rolled over. He rolled over too—as if he were going to do a split-S maneuver—where you drove over 180 degrees on your back and dove out. Instead of that, though, he rolled back out of the maneuver. I had started to give up on him, but when he rolled out, I twisted around and got right on his tail and got him. I thought I got him, I didn't flame him. I could tell he was a goner when I saw the burst of exhaust out of his stack, and then when he slipped off and fell. At that time, I turned around to my right to regain formation, but just as I did, I saw two Imperial Navy fighters coming at me. No sooner had I seen them, they opened fire on me. I was so scared! I didn't see anyone coming after me, so I figured I might make it. Still, I went over and got into the clouds and hid in there for a few seconds, then came back out.

As I began to climb out toward the rendezvous point, I saw four planes coming and thought. "Oh, hot dog, I've got help." As they got closer, I could see they had great big meatballs on their wings. That scared me again. I knew they hadn't seen me, they didn't see me! I finally got up close and they still hadn't seen me or come after me. I swung around into what we call a low side run and got on the lead plane. I put a proper lead on him and fired. I saw the engine cough and watched the plane fall off. I held the trigger down for as long as I could, which was really just about three or four seconds, but seemed much longer. I raked the next three planes as they were going by. Then I turned around and headed like crazy out of there. The three planes turned around and came after me, firing. I was going nearly wide open, so they didn't hit my plane, but threw a lot of tracers.

You can't see in back of your plane and I didn't dare turn. As a consequence, I couldn't tell what these three planes were doing. I saw something high above me. Immediately, I recognized it had gull wings—and there was only one thing with gull wings: a Corsair. So I called up and said to the Marine, "Corsair, Corsair, way above me, can you see me down here?" The answer came back, "I think I can see you. Make a left and right turn for identification." I did. He said, "I have you in sight. Let's go get them!"

We headed back the way we'd come and at about that time, I looked over and saw a Zero passing us on the right. The Zero pilots didn't see us. I signaled to the Marine, indicating a "bogey" on the right. He took one look and went after him. The Marine was going to get the first shots in, so he opened up, but I observed that he only had three guns firing. Within a second, he only had one gun firing. I reacted and headed in for a run, but he opened up again with that one gun and blew the Zero up.

*What were the most slugs you've gotten in your planes?*

The time the Imperial Navy pilots opened up their guns on me, I had seven or eight slugs. I was pretty fortunate that time. When I got back to the base, I noticed the slugs. In fact, one slug was in the armor plate behind me. It had not penetrated the armor plate. The slug would have hit me in the shoulder blades, if I hadn't had the armor.

*When you came back after the dogfight, what did you think?*

I was pretty chagrined. When I was debriefed, I was kind of shamefaced about it. It sounded like I had been pretty yellow, pretty chicken. It bothered me even though I remembered that Rickenbacker had told us about a time when he had a shot a plane down and when three other planes came after him, he ran like hell. He told us that that way he was able to fight another day. That was what I was planning to do.

*Don't you think you were being a little hard on yourself?*

One still feels shame-faced to run like hell. But during the debriefing when I was describing the corkscrew, some Corsair pilots came up behind me and were listening. One of them laughed like hell and said, "We thought you were a goner." I said, "I was diving to get the hell out of there." He said, "Well you did a good job. We couldn't have saved you—we got there too late." Then he smiled and said, "You'll be glad to know that a few seconds after you squirreled out of there, we blew those two Imperial Navy pilots out of the sky." I felt a lot better after that.

Another time, one of our planes was beginning to have trouble, a Zero on his tail. The pilot hollered for help. One of our pilots, a ministerial student named Buzz Carlson, heard the yell and before any of us could react, whipped his plane around and blew the Zero out of the sky. It may sound ridiculous to say you blew them out of the sky, but if you got close enough and fire tracers in the fuel tank, that's exactly what you did,

When Buzz got back to the base, he went out walking alone. It bothered his conscience that he'd killed somebody. Buzz used to get teased unmercifully. This one guy would pretend to be drunk and say he was going to shoot out the lights. He was supposed to be mad at Buzz. Sometimes they'd run Buzz right out of camp. This one time I didn't know they were kidding, and I said, "Hey, somebody get that crazy son-of-a bitch and put him to bed." I was going to confront him, when someone told me the guy was kidding. I thought it was cruel. Anyway, after Buzz shot that plane off that pilot's tail, he never got kidded again.

*Did you lose a lot of good friends?*

Yes. In this first little squad I was in, we had only eighteen pilots initially, and we lost five. In the next squad I was in, we had sixty-five pilots but lost seven. I say only seven, but we lost four out of seven to enemy action—two each on two different days. The loss that most affected me happened during the first cruise. We didn't have enough carriers to go around and our men were still hanging onto Guadalcanal. I got word that my friend Bill Bartlett and his Marine squadron were stationed at a camp nearby. I jumped in a jeep and rushed over to see him. His Skipper got a funny look on his face and said for me to sit down. He then told me my good friend had been killed in a midair collision. That really crushed me. That was the worst few days after that. I just wasn't worth anything. There is comradeship during combat tours that I have not found elsewhere.

*How did you remember Bill?*

We were competitors and friends. Without bragging too much, Bill and I were very good shots, and we had a friendly competition to see who could shoot the best. We got to where we were shooting twenty-something percent on target sleeves. One day I'd tell him I hit twenty-three percent, and then I'd hear from him the next that he hit twenty-eight percent. That was damn good shooting.

*What was average?*

I would depend what you were shooting. The big banners—six by thirty feet—were easy to hit, but I'm referring to percentages on a little sleeve—three or three and one-half feet by twenty, or so. People would say you were doing great if you got anything between eight and fourteen percent on the small target.

*How many kills did you have?*

I only had one and one-half verified kills, but I think I actually had three or four. Several Zeros looked as though they had been hit, but I didn't stick around to find out for sure. Those don't count as verified kills.

*Talk about flying Vella Lavella Island in the Solomons.*

We were camped there for a month or two. Once at Bougainville Island, the Japanese shelled us at three a.m. We had thought we were secure there, but it turned out that the enemy had carried a cannon all the way from the other side of the island and stuck it in a little cave. We woke to hear a shell whistling into camp. When that happened, Skipper told us to get a quick breakfast and prepare to fly to Treasury Island. As we were waiting our turns to take off, another shell came in and made a direct hit on a dive-bomber that was sitting on the end of the runway. I was just sitting there watching when there was a tremendous explosion. There was nothing left, just smoke and steam. At that point, our superiors told us not to go out and sit on the runway, but make a running takeoff, which we did. I had a tire go flat on takeoff, but I got out. When I reached Treasury Island, I let everyone else land first. When I finally touched down, I held off as long as I could, got on the right brake to equalize, and rolled in almost straight.

Once at Guadalcanal we heard about Japanese Betty Bombers coming. A night fighter flying a special "Black Cat" fighter plane, planned to go up and intercept the Bettys. The Bettys came across and laid a string of bombs near the runway. When the bombs hit, we rushed out of the Quonset huts and the compression of the bombs almost knocked us on our asses. The night fighter came up behind the Bettys and opened fire. He missed. The whole pilot group, as if one person was speaking said, "Lead him, lead him." The searchlights, radar controlled, came on; hitting the Betty's. The next time the night fighter opened fire, he was firing slightly ahead of the Betty, and he walked the bullets back into the Betty. The Betty flamed, caught fire and crashed. That one was right over the field. There were three Bettys. He chased another one and shot it down also. Boy, that night fighter was some hero around those parts after that. That was about the last time I saw the Japanese bomb Guadalcanal. We were there for about a month and a half. I don't think they bombed there after we left. We shortly moved to Segi Point, Vella Lavella, Bougainville, Treasury Island, and finally to Green Island, about twenty minutes from Rabaul.

*Who was the biggest character you knew in the South Pacific?*

Major "Pappy" Boyington was out there with us. I was a fledgling ensign then. Major Boyington loved his booze, so every time we got some beer he magically appeared. You would not refuse him any. But he was about as colorful a character as you'd ever want to see. One time Boyington had flown up high, ahead of his squadron. Some of squadron-mates took off and attempted to catch up with him. When they caught up and saw him, he had four Zeros attacking. One of his men called up and said, "Hang on, Skipper, we're coming." He answered, "Go find your own. I got these four here covered." I heard him say that.

*That's the stuff of legend. What was the best flying you ever did?*

After the Armistice was declared, the military decided to put a "Group Grope" out there, along with fourteen carriers. We were to fly over Tokyo, Japan, but as we got up in the air and headed out, the weather got worse. The clouds were low and the planes above us kept coming down. We were weaving back and forth. It was chaotic. Trying to get twelve to fourteen hundred planes to two thousand feet of airspace is not easy. Miraculously, we didn't lose a plane, but we arrived back on the carriers really sweating from that experience.

One time in Tokyo Bay in July 1945, we were flying back and forth above the clouds trying to spot the places in Tokyo for torpedo bombers and dive-bombers to bomb. We didn't know we were being fired on. The Japanese had some new smokeless powder that we couldn't see. All of a sudden, a pilot of a torpedo bomber called and said, "We've been hit. We are going to glide out over open water as far as we can." In that moment, the sky went crazy. All of the planes started weaving to avoid the flak. The planes dropped their bombs as best they could and headed out. Meanwhile, the TBM was gliding out toward open water. It got out as far as possible, but was still within range of shore batteries. Someone notified the PBT "Dumbo" to help. He was taxiing in, but still was about a mile away. Just then, a submarine surfaced near the TBM. Someone radioed from the sub telling the Dumbo to get out of there. The submarine picked the pilot up and was out of there in two minutes.

*You heard this on the radio, too?*

Yes. The squadron I was in was ordered to return to base, but I heard most of it. We were blocked off from some of the signal.

At Vella Lavella, we had what we called "Crash Hill." The pilots would go up there and watch when we knew a plane that was having trouble was coming in. One time we learned that a P-38 didn't have any flaps and was going to try to land. In the distance, we saw the plane coming. It was only about ten feet off the water. The propeller was digging into the coral, and all of a sudden, smoke was pouring out of the plane. We thought, "Oh, Lord he's going to burn!" He must have thought so too, because he was trying to climb out of the plane. He had slung his right leg out and was getting ready to put the parachute out. The plane must have been going eighty or ninety mph at this point, because he had slowed some. When he first landed, he was probably going two-hundred mph. Just about the time he was ready to jump, the shaft to the propeller broke and the propeller went right over his head. It disappeared so fast that he apparently decided to hop right back in the plane. It seemed he would rather face the smoke than hit the ground going that fast. The plane went three-quarters of the way down the runway and then slanted sideways. He was finally stopped when he climbed out of the cockpit and ran full tilt off the wing. Happily, he only got a sprained ankle out of it.

One time we were flying a mission and something went wrong with my plane. The batteries were pooping out and the gauges were dropping fast. I thought I might be losing fuel. I asked my skipper to let me run for the base. He released me, and I poured it on to get to the base. I was flying high, but as I came across the end of the runway, I feared I was running out of gas. I came across the end of the runway in a sideslip. In a sideslip the pitot-tube, which measures air flow and gives air speed, doesn't read right, so I had to disregard the air speed indicator. Moments before, when I kicked it out to sea and came over the end of the runway, I was doing one hundred twenty-five knots. When I tried to land, I touched once but bounced back thirty feet in the air, then went down again. At this point, I realized I was going too fast. I had only one to two hundred yards of runway left, when suddenly I looked to my right and saw a revetment slanting off at a thirty-degree angle. I thought that's my only chance. I pumped my brake three times and it turned me about ten degrees each time. I found myself going down the revetment at 55 knots. To my good fortune, it was clear and no airplanes came looming up. But that was a close one!

*Did you ever have a little prayer when something like that happened?*

No, but when those two planes that I described earlier opened up on me, I was repeating the 23rd Psalm. "The Lord is my Shepherd. I shall not want . . ."

*Did you ever get depressed?*

The toughest thing for me was during the second tour. I spent several months out at sea on the *Ticonderoga,* and that got pretty old. Once we rode out a typhoon—the huge one where three destroyers sank. We were actually right on the edge of the typhoon—out about as far as we could get from it. The waves were tremendous! They were running forty, fifty or sixty feet. All of a sudden, while we were refueling with this six-inch hose, we heard the ship's horn blare: boom, boom, boom—our warning about a big wave ahead. I was standing on the hanger deck and looked up. I described it as a hundred-foot wave. Near us was an old Canadian destroyer being refueled. The carrier rode up with the wave, but the little destroyer was caught nose-down, and it just disappeared—stacks, radar mast and all –for a few seconds. Then it came up shaking like a dog. I couldn't believe it was still there. There were three men on deck. All were injured, but they hung on. I've thought about that a lot of times. How the hell could they hang on? But they hung on and got out of it all right.

*So did you?*

Yes, I did.

*How would you like to be remembered?*

I was such an insignificant little cog that I don't care how I'm remembered.

*Those were the times, though, weren't they? You never get over that war experience.*

No, you wouldn't. In fact, everything is anti-climactic after that.

# Amelia's Way

## Amelia Reid

Amelia Reid's first ride took place when she was in high school, in the summer between her sophomore and junior years. Her hometown of Ord, Nebraska, did not even have an airport at the time. But it did have one of the youngest female pilots who had ever earned a commercial license, at the tender age of 18. Her name was Evelyn Sharp.

Born November 13, 1924 in Ord, Nebraska, Reid soloed in the fall of 1945. She earned her license in May of 1946 in Washington, D.C., and later flew for Ames Research. A consummate flight instructor, Amelia has taught more than 2,000 students to fly and has logged 56,000 hours herself. Until recently, her hobbies included roller-skating. "I was really good," she said. "I was doing turns and skating backwards when I took a really bad fall and hurt my back. I said, 'I don't think I'm going to do this anymore."

Before we interviewed, she told me she is smaller than she used to be. She claimed she had gone from 5'4" to 5'3"—but still stands tall as a pioneer in aviation.

Amelia died March 3, 2000, at the age of 76. After her death, one friend wanted to honor this remarkable woman and change the name of the street in front of her place to "Amelia's Way."

*What in particular do you remember about that ride with Evelyn Sharp?*

I remember going up for the ride and looking down. Everything looked so small. It impressed me. My parents knew her parents, and they gave me this ride. She landed in a grain field northwest of town. I was so impressed with it that it all I could think about was that ride. I started buying any books I could find on airplanes. At that time, there were a lot of books on the military planes, so I bought a lot of them and started reading up and studying the different planes. I was so interested in flying that I went to college where there was an air base.

*What was the name of the college?*

Kearney State, and I stayed in a dormitory and would go upstairs to get a suntan. While I did that, I could be watching the military airplanes fly over the dormitory. I got pretty good at identifying them because of all the books I had read.

*What did your classmates say? Did they think you were kind of "kooky"?*

They didn't say anything really. In college, I did a four-year program in three years and two summers. I didn't have a car at the time, but I did have a bicycle, and I used to bicycle out to a private strip that was really a grass field south of the college, about three miles away. There I would take the spending money I saved and spent it on lessons. My whole goal was to get a pilot's license. I took my flying lessons in a J-3 Cub.

*What do you remember about Ms. Sharp?*

She was five years ahead of me in high school and had taken flying lessons in a neighboring town, which was 70 miles west of our hometown. When she got her commercial license at age 18, she was barnstorming to pay off loans. She soloed when she was 16, got her private license when she was 17, and then a commercial license at 18. She went on to become a WASP in WWII.

*Did she give you any advice or tips?*

No, I didn't talk with her. The only thing I did was ride with her. But I never forgot it.

*What did you think of a female doing this? What were your feelings?*

I don't think I had any thoughts about this. My dad had a hardware store, so I was raised with tools and hardware, and I just thought nothing of it. The thing that always got me was that my mother used to say, "You're a girl, you can't do this." But I really wanted to do it, and she knew that she could never stop me.

*You were pretty strong-willed?*

Oh yes, I was very strong-willed. I still am! I just wanted to do something, and I did it. After I graduated from college with a teaching degree, I realized I would not take a job in the home state teaching school in my hometown. I didn't want to do that. When I graduated from college, my goal was to go somewhere I could learn to fly. I wanted to finish up my lessons and get my pilot's license.

*Tell me about your first solo.*

I remember my first solo took place on a field in Washington, D.C., that doesn't exist anymore. I used to hitchhike out to the place, which was called Congressional Field. It was a square grassy meadow. I soloed in a J-3 Cub, and I still remember bouncing the airplane around, landing and taking off.

*What was the best and worst advice you ever had in your early days of flying?*

I don't recall anyone giving me advice—or if I paid any attention to it, I don't know. I went through classes that taught us how to navigate, and since we had a woman instructor in the school, I never thought anything of it. The East Coast was very liberal at the time. If a girl wanted to do this, it didn't make any difference. When I moved out to California, I noticed things were a lot different – they thought I was crazy to be flying. Back on the East Coast, there were other girls learning to fly, and there was actually a woman instructor.

*Do you remember any of these other girls?*

I didn't really have time to associate. There was one girl—though I don't remember her name—who was working at the airport doing fabric work, and I really thought that was neat. I got a job at the airport taping joints, and nobody thought anything of it.

*Is there anything that shows just how obsessed you were with flying?*

I was so determined. I didn't have any transportation when I went out to Washington, D.C., so I had to get a job. I worked during the week, and my whole goal was to get a job and earn money to fly. That was the goal: to fly. I was renting a room in a house, and there were streetcars running out in front. I could take the streetcars out to the edge of the city, and then I would hitchhike. That's something you wouldn't think of doing now, but I did it then and didn't think anything of it. To look like a pilot, I bought myself a pair of khaki pants and a khaki shirt (like the guys wore in the military) as well as a leather jacket. I was that determined. Traveling salesmen would pick me up, and they'd say, "Hey, young man, where do you want to go?" and then they'd recognize me and say, "Oh, you're a girl!" I'd be out there at the airport every day.

I worked at two places in Washington. The first was in the Navy Department. I was only there about six months. I also spent six months with the Bureau of Standards. We did research and testing on ballistics, and we had a course where we fired a gun. The work I did there was really interesting.

*Then you were contributing to the war effort in a way, weren't you?*

Yeah, in a way. My whole goal was to get my pilot's license and then fly with the women in the service. But I got my pilot's license about six months after they closed down the WASP program.

*You must have been broken-hearted?*

Yes, because I wanted to fly in the war. I really wanted to be a fighter pilot!

*Did you really? Did you have dreams of that?*

Yes! I wanted to fly P-51s. Of course, I never got to join the WASP's. I was working for the Bureau of Standards at the time, and I wanted to transfer to NACA (National Advisory Committee for Aeronautics). They did research on WWII airplanes, and I thought that would be really interesting. I had a choice between transferring to the NACA on the East Coast or the one on the West Coast. Since my parents had moved from Nebraska to California, I thought I would transfer to California. I had gotten a degree in college in mathematics, so I got a job working at NACA working as a "computer"— as they called it in those days.

*You were a computer?*

Yes. They had desk calculators that I did equations on. I don't remember too much about the course there. The gun trajectory didn't really interest me. I was more interested in the flight characteristics of the military planes. I can remember what really intrigued me was when they would change the angle of attack of the wing. The plane would stall. They would change the design of the wing: the angle, thickness, curve, and it really interested me.

When I moved out to California, I continued to fly. Then I quit flying for about five years when I put my first husband through college because we didn't have any money. After the divorce, I went back to flying again. That was when I got interested in becoming a crop-duster pilot. My second husband was a crop-duster pilot, and that's what I really wanted to do. I flew his crop-dusters for some time, but never dusted any crops.

When I got pregnant, I lost my job at NACA because, in those days, there was no pregnancy leave. You had to quit and they wouldn't save your job for you. I loved flying, so I joined a flying club flying as a private pilot and just had fun.

After the baby was about eight months old, I decided I had to do something with myself. I became a flight instructor. I earned my teacher's credential when I got my bachelor's degree, so I thought why not go ahead and get my instructor's rating. I had been flying all along and had compiled about 525 hours by then. I worked part-time teaching and would still bring my baby boy out to the airport. So, I ended up raising my son out here.

*Where is he now? Does he fly?*

He flies for Northwest Airlines as a copilot on 747 jets.

*Any memories about when you learned your son's earliest flying?*

My son was born in May 1959, so it was a year later, in December 1960, that I got my instructor's rating and started teaching. I used to take him with me when I'd go out and fly cross county with students. I'd put him in the back seat. My son was with me all the time while I was out here teaching. He was just raised in the airport atmosphere. When kids were about three or four, parents would start teaching them the numbers and the alphabet, but he just wasn't interested in that. He just liked being around the airport. I knew he had to learn to read and learn his numbers so I bought him a model airplane kit. He was about three years old at the time, and the book was a very simple one with simple instructions, but I decided I wouldn't read them to him. He had to learn to read those himself. That's how he learned to read—by building model airplanes. After that, he could read instructions and started building all kinds of model airplane kits.

All the time, he was learning all this at the airport, and when he was 15, he said, "Mom, I'm going to be 16. I'd like to solo." He had decided on his own that he wanted to solo.

*That was kind of a proud moment for you, wasn't it?*

Very proud, yes.

*What did you do?*

I soloed him in five, tail dragger airplanes.

*You still teach in taildragger aircraft.*

Most people come to me because that's what they want to fly in. They know that's the best way to learn to fly because they learn more about the fundamentals of flying in those airplanes than they do in anything else. They learn to use more than three controls.

*What's your favorite taildragger?*

When anybody asks me that question, I always say, "The one I'm flying right now, the one I'm in."

*Did you ever get any grief when you had your little boy with you? Did anybody ever say to you: "Mothers shouldn't do that"?*

One time, my son and I were flying a taildragger—a Taylorcraft L-2 from Evansville, Indiana, back to California. I usually would follow the highway on the leg from Midland, to El Paso, Texas, because you've always got a place to land when you're on the highway. It was overcast that day, and my son was sitting in the back seat. The ceiling started coming down, and I made one attempt to turn around over the highway. I said, "Well, let's go back to the airport." We did and landed back at Pecos, Texas, and waited a couple hours, then we tried again. There was just no way to get through over the ridge because the clouds went right down to the highway. When I made the second turn-around, I was pretty low, and the cars that were driving down the highway started scattering when they saw the airplane. At that point, a Texas Ranger called the airport and complained that there was a low-flying, blue and white airplane. The airport manager told the ranger, "I have two airplanes there—a red and black Cessna and a lady with her little boy."

*Did anybody say anything like, "What are you doing taking that boy out, it's dangerous"?*

My son was 16, on May 13 and about three months after he had soloed, there was a fly-in for an air show. He said, "Mom, I'd like to go to the fly-in in Vancouver, Washington." I didn't think that sounded like a bad idea. I had already given him cross-country training, so he knew how to do all of that. He had had 150 hours of flying cross-country, but he didn't have his license yet. Here he was, 16 years old, student pilot, had just soloed, and I let him take my cub and go with a bunch of guys to this fly-in. The ladies at the airport were shocked. They said, "How come you're letting your son go all by himself clear to Vancouver?"

They couldn't believe it. I said, "Well, I wouldn't let him drive a car that far." I wouldn't even let him drive the car to Santa Cruz when he was that age! But, I'll let him take the airplane and go clear to Vancouver.

*What makes you a great teacher of flying?*

When I teach, I try to break down everything so simply that absolutely nothing can keep somebody from handling the controls. I tell my students to make it fun.

*What's the strangest thing you've had happen when you were teaching a student? You've had some unusual students I'm sure.*

I think one of the strangest things I've had was when a student who wanted to learn to fly in a Taylorcraft said to me, after his third lesson, "I want to rent it now, and go to Los Angeles with it." I thought, "There's no way!" He was dead serious he wanted to do that. But I refused and never saw him again after that.

*Have you ever run in to anybody that was just indefatigable, "I want to do it my way" maybe too strong-willed?*

Oh yes. I finally tell them, "All right, go ahead and do it your way." But when the student loses control of the plane and I have to straighten it out, they get very subdued.

One fellow right now taking lessons from me is an example. He is so strong-willed that I just sit there in the back seat and let him goof around and do what he wants with the airplane. I'll retrieve once in awhile, when he does something wrong. I figure he'll be out there messing around with the plane, and I don't like that. So sometimes I'll pull a little reversal where I just will not rent the airplane to the student.

*So you have to be strong-willed, too, don't you?*

Yes, very strong-willed. I had one student who wanted to teach himself to fly. He didn't want to listen to me, so I said okay, and I let him go as far as he could. When it came to the landing work, there was no way he could do that, or even perform the takeoffs, so I would retrieve the controls and finally he'd give in. After they scare themselves a little, they'll finally give in.

*Did you ever have any forced landings with students?*

I had an emergency landing one time with a student in a Taylorcraft. We had a hole in a piston, and we lost all the oil. That was in the day where there were a lot of fields, and so when I noticed our oil pressure had dropped down and temperature had gone up, I knew we had to make an emergency landing. We landed just fine in the field about five miles from the airport. We found a house and called the airport and they came out with some oil. We put the oil in the airplane, but I had the student drive back in the truck to the airport and I took off. I took it at 1800 RPMs. I thought that was enough to get me airborne in this craft. I flew it and landed on the taxiway at Napa airport. I had to fly under power-lines in order to make it, since I couldn't get any altitude. That student went on to become an airline pilot. He checked out on the 747.

*You must've come awful close to power-lines.*

Yes, I did. I had to go underneath them.

*Were you frightened afterwards?*

No, it didn't bother me. I just went ahead and did it. I had another exciting experience when I was by myself one time. I was up practicing spins, and the engine was idling a little slow, so I pulled the nose up and I closed the throttle all the way back, and all of a sudden, the prop just stopped rotating. It came to a dead stop. I was at 3,500 feet. Without any training, just from things that I had done hand-propping airplanes, I thought, "If I dive this Taylorcraft straight down, the red line is 130mph. I figured with air pushing on that prop, maybe it will start." By golly it did! It started and I got back to the airport!

Back in the 70s, I was trying to teach myself aerobatics. At the time I was having trouble doing hammerhead turns, so I talked to some guys about how to do them. They told me to put the plane in a very flat spin. There was a funny sensation. There you were, hanging on the belt rotating upside down. I'd never experienced that before. I'd gone to seminars, and they said, "If you ever get a flat spin, don't fight it – just relax and cut the power off." I had full power and was rotating like crazy, so I cut the power, let go of the control, and then I came right out of it into a regular spin.

*You sound like a kind of cool customer. Have you ever been frightened?*

I don't think so. I don't ever recall that. I just do the job.

*A lot of times people talk about coming back from something scary and sitting there with their knees shaking.*

I've never had that happen.

*Do you have a personal idol?*

Evelyn Sharp is the one I always admired. Otherwise, I don't think I think of anybody as a hero.

*You didn't think much of Amelia Earhart?*

I didn't think much of her flying. She had the money and didn't do it the hard way. I think I would be more interested in somebody who struggled.

*How do you account for that kind of attitude?*

I think it's from my family. My parents are both from Czechoslovakia, and they have struggled very hard to get anywhere. They worked hard all their lives. I think that's where it all comes from.

*Were they as unrelenting as you are?*

Probably. My mother didn't go along with my flying, and my father really didn't care.

*When did they give you some sign that they were proud of you?*

My dad didn't live to see all these things that I'm doing now. He just didn't live that long, but I think he would have been proud. I think they both were. They rode with me a few times in an airplane. They didn't stop me from doing anything. If I wanted to do it, it was okay. That was kind of their attitude: "If it's what she wants to do, let her do it."

*What traits do you get from your father and what traits do you get from your mother?*

That's a good question. I don't know. They both worked hard. My father ran a hardware store, so I think I might get my business sense from him. But they were both proud of their kids whatever they did. They just accepted whatever the three of us did.

*Do you recall a student that you are most proud of?*

Yes, Sean Tucker. When he came to me, he was just frightened of airplanes. He had been a parachute jumper, but for some reason he decided to learn to fly. He was dead serious, even though he was scared. He thought maybe he could learn aerobatics. I took him up, and I started teaching him aerobatics. He became so intrigued with it -- especially when he found out that the airplane wouldn't come apart! He got so excited and exuberant! He loved it. The funny thing was that he got so interested he wanted to learn more stuff. So behind my back, he decided to go out and practice slow-rolls, and he broke the door in half. I grounded him! He still talks about how I grounded him.

*You say he broke the door in half?*

He was trying to do a slow-roll, and the door started vibrating, and the vibration broke it in half!

*Who else are you especially proud of as a student?*

I was proud of my son—very proud of him. He flies so many different types of airplanes that I have never flown. I keep flying the school planes, but he gets into everything.

Once I had a lady who came to me, and she wanted to travel around the world, but she was absolutely frightened of airplanes. She would sit there and chew her fingernails off. Her roommates suggested that maybe she ought to learn to fly to get over her fear. She was absolutely scared to death. I remember giving her lessons, and she would come back after a flight, looking so scared she was just white.

Well, I like to analyze my students and then give them the kind of attention they need. Some of them you can yell and scream at, and some of them you can't – you have to be very gentle. I remember making some remark about flying: "Well, maybe girls shouldn't be hand-propping airplanes. Maybe we should just let the guys prop the airplanes." That really upset her because she wanted to be independent and do it herself, just like I did. She was bound and determined that she was going to learn to fly, and she got over her fear. Later on, she got all her ratings, and became a flight instructor. I was very proud of that.

*What's the wildest stereotype you believe someone's had about women flying? Some statement or something you've heard?*

Generally, most of the men say that the women have a lighter touch on the controls, but that's not always true.

*So you've got the gentle touch, huh?*

Yeah, that's what people say about my aerobatics. When I take someone up and perform aerobatic rides for someone, they say, "I can't believe it. We didn't feel anything," because I can do it so gently for them. That's what I try to do, and that's how I try to teach aerobatics.

One day when I was flying aerobatics, I was trying to do a multiple-turn spin. I went in to the first turn, the second turn, the third turn, the fourth turn. During the flight I had a student in the plane. All of a sudden, I started a spin recovery—like they teach it— and the airplane wouldn't stop spinning. I cut the power all the way back, and it just kept rotating. Finally, it slowed down after we lost some altitude, and we came out of the spin. I got to thinking about this: "Why did this airplane do this?" It was a few weeks later that one of my competitors here at the airport had an aerobatic pilot out doing the multiple-turn spins. When he went into the last turn, he didn't recover in time and flew right into the mountain and killed himself and his pilot. There was a lot of discussion about what caused this. Word of this got back to the factory, and they did some spin testing.

Some FAA certified airplanes are spinnable and are able to recover in one turn by letting up control, but they didn't do any tests for multiple-turn spins. So the factory tested the airplane and found out that, sure enough, this particular airplane would not recover. To this day, when I teach students, I restrict pilots to only two turns. You don't need to do any more than two, and we won't have any problem with recovery.

*What's your favorite aerobatic move?*

I like the rolls in a Cessna Aerobat. It's just beautiful. I do one that's called an avalanche. I really like that one, where I do a snap roll on top of a loop. But, then again, I like them all.

*Why do you like that avalanche?*

Because it's fun and you tumble in it.

*Was there ever a show where everything clicked so perfectly that it was an absolute dream?*

I would say that this last show that I flew probably fits that. I had had so much practice that everything just clicked. I did four routines in Santa Rosa, and every single one of them was identical. I think the reason I felt my performance must have clicked really well is because after I landed and got out of the airplane, and lots of people came up to me and seemed to be in awe. In fact, I couldn't walk three steps before somebody would come up to me. But, when I think about it, I realize it was just a lot of hard practice. I practiced very hard for that routine. There were just two of us flying in that show, and I wanted to be perfect. On the other hand, I don't like to think that I'm doing something perfect, because then I think I can't get any better.

*What were some of the maneuvers you did?*

I did a three-turn spin, I did loops, and I did a Humpty Dumpty.

*Do you do Immelmanns?*

I do. I also do hammerhead turns, snap-rolls.

*If there an aerobatic move that best matches your personality?*

Yes, it's the one that everybody likes: the Butterfly. It's not an aerobatic maneuver, but it's my low-altitude, 40 degrees of flat turnarounds.

*How does that match your personality?*

Because I'm down low on the ground and my wingtip is almost scraping the ground on the runway. Then I make a turn and I peel on up at 40 degrees so the airplane is just about hanging on the prop, and then I dive back down toward the runway. I call the maneuver a Butterfly. The people are very impressed with that routine. I think I enjoy that one the best.

*So that matches your personality?*

I don't know why exactly, probably because I'm up and down all the time!

*How many students do you generally have in a day?*

I run between five and six flights a day. That's five or six hours.

*Where do you get your energy?*

I'm very careful what I eat.

*Is that right? Tell me what a little about your diet.*

I eat oatmeal for breakfast, grapefruit, orange juice, and milk. I have a banana snack, broccoli snack, carrot snack, apple snack. A cheese Swiss sandwich on rye bread. I love Turkish dried apricots. They're really good. That's the kind of stuff I eat. I try to avoid anything greasy–hamburgers and such. I'm not too heavy on meat.

*What are you proudest of generally, in terms of what you've achieved?*

I'm proudest of doing aerobatics, which I've always loved. I'm proud of being able to do that. I think I'm proudest of being able to teach people to fly, of taking folks who just know nothing, and making pilots out of them. I think that gives me a really big thrill. Oh, everybody thinks I was named after Amelia Earhart. When my mom and dad were courting, they were in a play in which my mom's character's name was Amelia. Of course, it was a Czechoslovakian play, because they were from Czechoslovakia. The Czechoslovakian translation for Amelia is Amarka. So my birth certificate says Amelia on it. I was named after a part that my mother had in the play! What embarrasses me is the actual part she had in the play. She was a maid! She wore a little apron and dusted furniture.

*You decided against being "Amelia the Maid?"*

I didn't want to be "Amelia the Maid." That's probably why I never do housekeeping!

*How helpful were men to you early on?*

Back on the East Coast, my instructor was a man, and he was a good instructor. Teaching a woman was no different from teaching a guy to fly. The men gave me jobs, and they worked with me at the airports. I worked at the airport. I hitchhiked out there, and I'd work all day Saturday and all day Sunday for one hour of flying time. Can you believe it? There was a guy out at the airport that let me drive his car back. He had a brand new Taylorcraft that he'd bought, and he had to go to Cleveland, to drive his car back, and he let me drive his car back so he could fly the airplane. I was driving in unfamiliar country that I'd never driven through before—and he let me drive his car. It worked out fine.

*Do you consider yourself a brave person?*

I just consider myself doing what I want to do. But I don't necessarily think of it that way. I don't let someone stop me from doing what I want to do. They tried to stop me from building my 60 by 100 foot building back in 1967. The Reid family owned the airport, but they sold it in 1960 to the County of Santa Clara. The plan was that I could operate there for three years and then, after the three years, the county was going to tear down all the old buildings. At the time, I was operating out of a trailer, and everybody had to get a contract and a lease with the county in order to build. The county came to me when I was in my trailer.

I'll never forget the day they came to me after hearing I wanted to build a building, saying, "Women don't do that. Women do not build buildings. Why don't you lease some space from somebody else who's building a building?" I said, "No, I want to build my own building." So I hired a lawyer. That's how determined I was.

One of my students was a contractor, and we got together. We got extensions month to month from the county to continue operating until my building was built. I took a loan, mortgaged my house, and with my retirement money that I hadn't used, I had enough money for the building with the hanger and everything. It cost $12,000 to put the hanger up. It's a four-beam building, and I got two-and-a-half acres of property that I paved and I got the whole job done for $37,000. That was back in 1967, and we used 30 years up on the lease and still have 20 years left.

*Do you have 20 years left?*

That's going to be a long time isn't it! I don't know!

*Do you ever get tired?*

When I get sick, I do. I usually go to bed about two o'clock, and get about six hours of sleep.

*And then to think you worked a full day. Boy, is it the genes, or the Turkish apricots? This is uncanny, though. Do you get any comments from senior citizens?*

You should have heard them at the air show when they saw me fly. All the seniors come. The senior citizens just can't believe I'm still doing it. They say, "Don't you get dizzy?" They just wonder how I do it.

*Is there another example of your business savvy that has really worked for you?*

I think that it helped in college when I studied typing and I learned how to use an adding machine. I experienced using an adding machine while I was working at NACA. I majored in math, and I think my mathematical background has really helped me a lot.

*Give me about a couple of pivotal points in your life.*

I think learning to fly was a very pivotal point. That was very important to me. Raising a son was very important too, and that he's successful. The flying is what I do all the time.

*How many hours do you fly in a year?*

I have flown between 1,000 and 1,200 hours a year, except for the five years the first time I was in college.

*You might be going for a world record one of these years!*

I don't know! I don't think about world records or anything like that. There's no goal to achieve anything. It's just what I wanted to do.

*What do you tell young women today who are interested in flight?*

If this is what you want to do, do it!

*How would you like to be remembered?*

I don't know. As a crazy lady, I guess..

*What did you think this year when you saw Sean Tucker at the Oshkosh Fly-In? Did he confide in you?*

We just visited a little bit. He was really busy.

*Are there any old-timers you thought were pretty amazing?*

My ex-husband was a tremendous pilot. I thought he was one of the best natural feel/field pilots that I've ever seen in my life. He could just wear an airplane. That's kind of the thing I like, is somebody who's such a natural at flying. We didn't stay married but I always admired his skills.

# Combat Over Japan

## Philip Igou

When Philip Igou joined the Air Corps, he didn't have any idea of what the training would entail. His parents decided they would transport him and a friend up to Maxwell Field, Montgomery, Alabama. Before they left, Philip packed two essential items: his tennis racket and golf clubs. When he climbed out of the car at Maxwell, a voice called to snap to attention. Surprised, Philip and his friend snapped to attention. Then the voice said, "What are you going to do with that tennis racket?" There was a pause. "You can send that stuff back." As a consequence, the racket and clubs made their way home and, what was worse, the two young men never saw a day off after that.

Philip had earned his pilot's license at the University of Florida through Civilian Pilot Training Program (CPT) in 1940. He attended basic training in Greenville, Mississippi, and graduated advanced training from Columbus Air Force Base in Columbus, Mississippi. Ultimately, this lean, easy-going gentleman flew missions in the Pacific bombing Japan in B-29s.

*Was training really rough?*

As soon as we checked in, the military barber cut our hair and then we were issued uniforms. Before we knew it, we were out in the field. I had basic formation at military school, but some had not had the experience. There, we actually had rifles with bayonets on them. We had one awkward character we trained with who kept cutting people with his bayonet. I had held a private pilot license before joining. On the first day of PT-13 training, an officer said, "Those of you who've had previous training, step forward." Some were smart enough not to step forward, but I did. Then the officer said, "Assume one position." "What do you mean?" I asked. Down there and put your arms out like wings. Then they ran you around like that. Talk about humiliation.

*Remember any tough decision. Or any regrets?*

I had a cadet I trained, and I regret it today. He was a student I called "Jittery Jake." He was a white-knuckle pilot. I used to try to calm him down. Ultimately, I wrote Jake up as "doubtful," but at the time, they needed pilots, and so they let him graduate. I remember I wrote a short information sheet on him, which said, "Under no circumstances should you put Jake into a single pilot situation, Like an AT-20. He does not have the temperament to go." Unfortunately, the military sent him down to Kissimee in an AT-20 plane, and he crashed and killed the crew.

*How did you feel?*

I didn't know it until later on. You think you should have been stricter, but they needed pilots and he met the criteria of being able to take off and fly.

*Any other regrets?*

We flew B-29s into combat over Japan and had been bombing at thirty-thousand feet. At that height our accuracy was missing, so Curtis Le May, who was in charge of the 58th Bomb Wing, came in and talked to us. We were in this Quonset hut. He told us that our next mission would be flown at night at seven-thousand feet. People just laughed, thinking it was a joke, but it turned out that it was a smart way to win the war.

At night you took off at one-minute intervals and dropped incendiary bombs in clusters, which hit and burned. Late in the war, Japan didn't have night fighters. Anyway, when we flew on a mission, we'd lose maybe three or four out of five hundred bombers each night. Sadly, though, one of my bombardiers, a nineteen-year-old kid named Tom Welles, was put on another flight. One night out of five-hundred planes, his was the only one that didn't come back.

*Give me a memorable mission.*

We were one of the last B-29s to arrive at this target, but by the time we got there the whole town was absolutely in flame. It was turmoil and fire below us. In the plane, it felt like a maelstrom of fire and updraft. So, I told the bombardier that there was no need to go into the middle of this, because the whole town was on fire. Distantly, you could see the town itself was burning. Why not go to the outskirts and spread the fire on out from there? This episode shows the difference maybe between a twenty-five-year-old, which I was, and a nineteen-year-old. Anyway, this young bombardier looks at me and says, "What's the matter, you chicken, Captain?" In my mind I was thinking, "I know he was instructed to hit the middle of town, but in my judgment, it would be better to hit the edge." But you have twelve-person crew, and you can imagine the crew coming back and telling the story. So I thought, "All right, I'll do it. I'll hit the middle." I said, "Hold on." Sure enough, when we flew over the town, the turbulence started throwing our plane around. It broke the chair my radar operator was sitting in, throwing him across the cabin. Yielding to my bombardier's desire for pinpoint accuracy could have been a fatal mistake. At that time, the fire was up to five thousand feet and covered a mile wide.

When we landed at Iwo Jima and started taxiing off, I told my copilot, "Look at that B-29. Look at the wrinkles on it." You could see the bomb bay doors were wrapped around the side. There were wrinkles all the way through the whole fuselage. I went over to the pilot of that plane and said, "What in the hell happened to you?" He said, "The fire threw our airplane upside down, and we started going down at about three-hundred mph. Both the copilot and I were pulling back on the stick with all we could muster. We got right over the treetops and pulled out of the dive." Strangely enough, we had flown over the same place he did, but the currents brought his plane all the way over. In retrospect, my first decision not to go was right, but I could imagine hearing the words, "My pilot wouldn't go."

Another night mission the same week my top gunner looked up above our plane, and said, "Captain, there's a B-29 above us. He's getting ready to drop his bombs. He's right over us." Again, we were at about seven to eight-thousand feet. I asked the bombardier, "How long before we drop?" He said, "Four or five seconds." I said, "Hurry it up. As soon as you drop, we'll get out of here." Do you get out of there before you reach your destination and drop your bombs? What does one do? Once again, discipline took over. "We'll wait the five seconds and drop and get out of here," I said. Just as we did and turned, down come the bombs right where we vacated.

*Ever become religious before you flew?*

Before every mission, I went to see the chaplain who held a brief service in the chapel. We had a down-to-earth chaplain, and one night mission he decided to fly with one of the crews. Before we would go, we would race our engines and apply full brakes. Then when you got released, which was at one-minute intervals, you'd take off and go out over the ocean. As we did this time, we could see a big ball of flames below us. The plane in front of us had crashed on takeoff. It was so black out there, that it was difficult to see. In fact, you had the bombardier looking to help you clear the trees. You could barely see the trees, or even see the ocean below you. I had gotten an eerie feeling on takeoff, but none of us knew who had crashed until we got back and found out it was our commanding officer, Colonel Watson, who had crashed, along with the chaplain.

On another mission at Tinian, this pilot named Hap Arnold was leading and here came this kamikaze in. The kamikaze hit the nose of his B-29, and that was the last of him and his crew. For some reason, the military took five B-29s crews and sent us to India after training. When we got our orders, I said, "Why? The 29 crews in India are packing up to leave over there and go to Tinian. They won't be there for but two or three months." It was case of orders getting screwed up. So we were the last of the B-29 pilots there.

*Did you bomb Burma?*

Yes, one time we lost an engine on a Burma mission. When you do that, you slowdown, which makes it impossible to stay in formation. At that point, we were flying over Burma I asked the navigator to pick an alternate target. The maps are supposed to give you all the alternate targets. A few minutes later, the navigator said, "I've got one, Captain. It's at 245 degrees." So we peeled off at 245 degrees. "How long?" I asked. He said, "It won't take us but about fifteen minutes." Soon we zeroed in and he said, "There it is." It was a Japanese airfield. At the moment we didn't know what to expect. We flew over and dropped our bombs on the airfield and barracks and headed back. We found out later the field was an abandoned Japanese field and that the British had moved in. The British were so angry that about a week later they fired a warning shot at one of our B-29s. It turns out the government had given us an old map.

*Did you do much damage?*

No, just some damage to the runways. Luckily, there was only a skeleton crew of British there.

On Tinian Island, the quarters were Quonset huts, and they were hot and uncomfortable. Bunks were tightly stretched canvas, which isn't too comfortable either. Before I went overseas, I had wrapped about ten or twelve bottles of booze up and hid them in my trunk. Then we went to India, and the booze never reached me there. Finally, it arrived in Tinian. There, booze was so scarce that a bottle sold for forty dollars. The Seabees were a creative group, who could find and ferret out any material you wanted. I had them make me a nice-sized bed made of plywood. They cut inner tubes into strips and wove them together for the support. Everybody was envious. "How'd you get that?" they'd ask. I didn't tell them, but I got it made for a bottle of booze.

*Any stories regarding your crew?*

My nickname came out of this situation. Let me explain. All crews had engineers onboard. I had an engineer named Grant. On our flights, you know that you've got four huge engines and a load of bombs and gas, etc. That amounted to considerable weight. As one of our newer crew members, my engineer said to me one day, "Captain, I think we ought to land at Iwo Jima because I'm doubtful about fuel," so on the first two missions over to Japan, we stopped at Iwo. On the way back after that happened, the other pilots came up to us and said, "What's going on? Why'd you stop? You've got the same amount of fuel as we do." I said, "We've got a different flight engineer." So a couple of these guys kept needling me. Then we landed a third time after a flight. That time we had an engine shot out over Japan and had to land in Iwo Jima again. Of course, that was a legitimate reason, but it marked the third landing in Iwo Jima in one week.

On the next mission the pilots and crews were in this hut awaiting destinations. They had a screen set up for one-thousand pilots and crews. At last, this fellow drew back the curtain and showed the destinations. He stood there with this giant pointer and said, "We'll take off from North Field in Tinian, and our first checkpoint will be Igou Jima." On the screen, he had written Igou, instead of Iwo. Everybody laughed. I took a lot of hell about that, but I had my nickname from then on: "Igou Jima."

# Korea, Vietnam, U-2

## Joe Jackson

He is a Medal of Honor recipient, to be sure, for his daring rescue in Vietnam. But Joe Jackson is an old flier who has seen a lot of action beyond the celebrated experiences.

Joe Jackson was born in 1923 in a farm house in Heard County, Georgia. His father was a country school teacher; his mother a housewife. He led a rough and tumble life as the youngest of seven boys and jests about how they had to burn the school house down to get him out. Upon graduation from high school, he planned on becoming an airplane mechanic, and went to a night school conducted by Delta Airlines. After he got through with that, he started looking for a job, but couldn't find one. He went back to the school and queried the instructor on why he couldn't get a job. "The woods are full of people who've had years in airplane mechanics and they can't get jobs," he was told. "What do you recommend?" Joe asked him. "I recommend you get in the Army Air Corps and get three years experience as an airplane mechanic." Thus started an illustrious flying career that involved WWII, the Korean War, U-2's, and Vietnam – to name a few parts of Joe Jackson's remarkable flying career.

Joe Jackson flew P-40A and P-63A in pursuit curve attacks to train gunners on B-17's and B-24's. He was scheduled to go in the Pacific and fly bombing missions to Japan, but the war ended. He later was sent to the European occupational forces and stayed there for a little over a year. Upon returning to the U.S., he joined the 27th Fighter Wing in Kearney, Nebraska. He attended the U.S.A.F. Instrument Pilot's Training School and found himself flying F-82's, an unusual plane with twin fuselages, one wing. Soon he had transitioned from the F-82 to the F-84 – a jet fighter. In 1950, he made one of the early missions flying fighters across the Atlantic, which he describes vividly in the interview. He flew 107 missions in Korea, following which he took command of the 467th fighter squadron at Turner Air Force Base. The wing was deactivated to form the 4080th Strategic Reconnaissance wing, which had the B-57's and U-2's. He was an early U-2 pilot. The service in Vietnam followed and, of course, his incredible 270 degree descent he made in his Medal of Honor rescue.

*Did you always want to fly?*

It was my objective when I was a kid. It really sort of came to me when I was in my teens around nineteen years old. When I was a kid, I guess I was one of the few people in the world that knew that I wanted to work with airplanes. I knew that since I was about five years old, but I never had aspirations of being a pilot—I wanted to be a mechanic—that was my original objective. And it was in 1942—after the war had started and I was flying as a flight engineer on a B-25—that an incident occurred. When one engine caught fire, the pilot turned around and said, "What do I do?" And so I said, "You feather it." You know, there's a procedure you go through and the first thing you do is to stop the propeller and it's called "feathering it" because the propeller is turned with the wind. And when I did that—or when he did that—the fire went out. The next thing that would have been done was to pull the fire extinguisher, but he didn't do that because the fire went out. I got to thinking—I'm just a young kid—a Pvt. 1st class with an air mechanic's rating—that if I know what to do and he doesn't know what to do—I'm as smart as he is—I can be a pilot.

I couldn't do it immediately because I didn't have a college education at that time. Two years of college was required. A buddy of mine who worked in the aviation cadet office called and said, "Hey! You still want to go to flying school?" I said, "Sure." He said, "Well, they got a new test. You can take this test and, if you pass it, you can go to Aviation Cadets without a college degree." I went over and took the test and passed it and went on to flight training in 1942.

*What do you remember about the training?*

Back in those days the aviation cadets got a lot of hazing. Hazing was very common then and people accepted it. It was just a form of teaching discipline and obedience to seniors, and that sort of thing. I'm sure there were a lot of anecdotes about that, but I don't remember them at all. I do remember one guy when we were out in the first phase of training, which is called primary flight training. We were shooting landings over at a little dirt strip in Tennessee, and this guy, in his final approach, ran into a tree and knocked a big hole in the wing. The airplane was made of plywood, so it knocked a big hole in it, but he managed to go on and land it. When he got back we were really razzing and hazing him about it and really giving him the needle. So he says, "What are you guys so upset for, I only hit a tree?" It was just a few things like that, but it was a lot of fun.

You know, we lost about 50% of our class in cadet training, that many washed out. They didn't complete the training so it was always a sad day to see them go.

*Do you remember anyone who picked on you?*

Oh ya! I was the little guy in cadet training. In fact, I probably weighed about 120 pounds. There was this great big guy, whom I really dearly loved. His name was Julius D.C. Anderson, commonly known as "Direct Current" Anderson, and he was always picking on me because I was a little guy and he was the big guy. But that was all in fun. Other than that I can't think of anybody that picked on me.

*How did people respond to you? You came from the Army, right?*

Well, of course, back in those days the Air Force was part of the Army. It was called the Army Air Corps. And the Air Force started as part of the Army. It was called the Aviation Branch of the Signal Corps. And it wasn't until 1947 that the Air Force became a separate service. Up until that time it was part of the Army. So, yes, I was in the Army Air Corps.

*Do you remember your first dog fight?*

You can never forget those things, I suppose. It was in the Korean War—I didn't get into combat in WWII because I was in the Training Command. Once I graduated from flying school and was commissioned, I was in the Training Command and was teaching serial gunnery down in Florida. One event that was amusing—I didn't think it was amusing at the time it happened when I was a cadet and was going through the final phase of advance training. We had guns mounted in the T-6's, which was the standard advanced trainer at the time for single-engine pilots. I got sent out one morning to fire ground gunnery. It was a beautiful morning and everything was just perfect. I came back after shooting my gunnery. About 30 minutes later my instructor came in—he was a captain and he called my name, "Jackson!" "Yessir," I said. It was common to stand up and hop to attention and say, "Yessir." He said, "I'm going to have you court-

martialed." I said, "Sir, may I ask the charges?" He said, "Yes, I sent you out this morning with a hundred rounds of ammunition to shoot at a target and you only got 94 hits – what the hell did you do with the other six?" That was a bigger joke than he thought because that was lucky. A lot of luck was involved on the flight—the weather was nice; the air was calm and the airplane had just recently been bore sighted, so it flew true—everything just worked out beautifully.

*What about your first combat missions?*

The first combat missions were mostly ground support and interdiction missions. With ground support you go out and you provide support for the foot soldiers, who are fighting. You shoot up targets they mark for you with artillery rounds or panels they direct you to by radio. Interdiction is where you go up past the front lines and try to intercept the supplies and destroy them before they reach the front lines—to try to out the enemies' ability to support himself. I did interdiction for quite a while, but my first aerial combat was, I think, the biggest jet-to-jet battle up to that time—that occurred in the Korean War.

It was early January and there had been some engagements jet-to-jet. I was flying F-84's at that time and had an early morning mission. The mission was to go up into North Korea and destroy a train that was sitting on the siding. Well, I had been over there long enough to know that the North Koreans didn't park trains on the sidings, especially when it had a bit of steam going up in the air from it and you could see it for one 100 miles. I said, "Hey guys, this looks kind of fishy. I don't think this is quite right." Then we got to talking about aerial combat. And, after the discussion, I decided that this was a trap so we should be especially careful. I had a policy that I would designate other people to lead the flight and I would fly wing positions to see how they were doing. I had another guy that was leading the flight that day. I was flying number four position on a relatively new guy. I was checking him out as what we call element lead. That is, lead for two airplanes. I had my deputy flight commander flying wing man on the guy that was leading the flight—it to observe him and see how he was doing. Anyway, we went on our way up north and found the train—we were loaded down with bombs and rockets. The sky was clear—it was a beautiful day in January of 1951. There were four of us. We spread out, taking spacing to attack this train. The guy leading the flight on this day was Pendleton and he looked back and called out, "Ohhh! There are swept wings in the sun!" At that point we jettisoned our ordnance that we had on our wings -- the rockets and bombs. We just toggled them off armed so that when they did hit they would do some damage. There were four of us at low altitude, low speed, and we were being attacked by sixteen, MIG15s, from high altitude. They were at higher speed and could out maneuver and out climb us.

We headed south and drew up a defensive formation. They kept pounding away at us—we kept evading them. It seemed like it went on forever. Finally, we got in the vicinity of the North Korean capital of Pyongyang. At that time they broke off and went home. We didn't inflict any damage at all, and we were lucky to get away. We were so badly outnumbered, outgunned and out-maneuvered.

*What techniques did you use to escape?*

It was the standard defensive tactics. If somebody attacks one of the elements, then the other element comes to his defense. That's the way we did it. It was standard procedure. We call it the scissors maneuver.

*It seems almost inconceivable that you escaped, being as you were so outnumbered.*

Well, we had a good flight. It was good teamwork and everyone knew what they were supposed to do. We did it and we were able to do it precisely. It doesn't mean we didn't get shot at. We got shot at a heck of a lot. We could see those tracers going by us almost every pass they made, but we were able to evade them and get away.

*How long do you think the whole confrontation took?*

Probably six months. Actually, it took maybe ten or fifteen minutes. But it seemed like a terrible long time.

*Do you have a favorite plane?*

Oh, I think everybody who ever flew it would say it was the B-51. That was in the propeller days. It was maneuverable, responsive, flew well, it was a good airplane. It was really racy, but it was just a super-duper airplane. It had everything you wanted.

*You had to get used to some odd-ball characteristics.*

No, it was a pretty straightforward airplane, I think. Maybe that's why everybody liked it. The wing design was a little different, which made it a lot faster than most airplanes up until that day. That was the first of what they call the laminar-flow wings—had a definitely different shape, a good, honest, nice-flying airplane.

*Didn't you receive the Distinguished Flying Cross as well as the Medal of Honor?*

Yes, I have been awarded two DFCs. On the first one we were attacking a supply dump that was up North in Korea, and the second Distinguished Flying Cross was for the re-supply of a mission down in Three-core in Vietnam. I don't remember the details very well.

*What about the Medal of Honor?*

Well, it was on May 12, 1968, right at the tail end of the Tet Offensive. On this specific day I was scheduled for a flight check. You know, even in combat you've got to take a flight check twice a year to see if you are qualified to fly. As ridiculous as it seems, that's the truth. So it was my day for a flight check. The flight check officer was flying as co-pilot and his name was Jessie Campbell. We took off with a mixed load of supplies and cargo, flying up and down the coast, up in I-core and then to Dong Ha, which is a little airfield, just south of the DMZ. We stopped at several airfields on the way up and picked up passengers and supplies to bring on the way back. Then we went South of Da Nang, to a place called Chu Li, which was a Marine base. But when we checked in to the Airlift Control Element, they told us to report back to Da Nang immediately.

*What time was that?*

That was sometime around noon on May 12. We went back in and they started giving us an intelligence briefing. We knew from that, that something was really hot! They briefed us about Kham Duc. The significance of Kham Duc was that it was about fifty miles west of Chu Li, right on the Laoian border. The reason for it being there was so they could observe Route 14, a supply route coming down just inside the Laotian border. It was the main re-supply route to the south. They had a decent runway, something in the range of 5,000 feet, with a hard surface, which was in itself, unusual. It was surrounded by hills. Because of its strategic location, the North Vietnamese decided they were going to eliminate it. They had surrounded the airfield with some four to six thousand No. Vietnamese regulars. The outpost at Ngok Tavak began to get into some skirmishes early in May. One day, in a skirmish one of the NVA was captured and he said they had the objective to take over the camp. Shortly after that, I guess, the camp came under heavy attack and they decided to withdraw. But en route back to Camp Duc, they got ambushed so had to go back to Ngok Tavak. Later, they withdrew to the south and were picked up by helicopters. They were brought back into camp. About the tenth of May the camp came under heavy attack. Because of the intelligence reports, they decided to reinforce with a battalion of U.S. Army, which they did under the command of Lieutenant Colonel Nelson.

He got into the camp, set up perimeter defenses and had about seven outposts to observe what was going on and to report back. During the night of the 10th and 11th of May, these outposts came under attack, and all seven of them were overrun. Nearly everybody was either killed or captured by the North Vietnamese. There were a few soldiers that managed to get back to the main camp.

*The main camp was under heavy attack.*

Yes, on the day of the 11th it came under very heavy attack from all directions. The North Vietnamese commanded all the high ground. They were dropping in rocket and mortar fire around the camp. They were really blasting it. That night, General Westmoreland decided what to do was to evacuate everyone from the camp: it could not be held. It was in such a position that it was not defensible. There was no way to get through by road, the camp was surrounded by mountains or big hills, and the North Vietnamese commanded those hills.

*So they decided to bring you in?*

Well, the next morning, they decided to evacuate and this was the morning I was to take the flight check. The first airplane that came in was reconfigured for re-supply, instead of evacuation, and had all these supplies on board. As soon as the airplane landed, it was immediately filled up with the dependents of South Vietnamese dependents off the airplane so they could get out of there. They finally made a pass down the runway, and they couldn't take off because they couldn't get up enough flying speed. They pulled back into the ramp, the dependents then got off, they unloaded the cargo. Meanwhile, they got out and were cutting the tire off with fire axes and bayonets to reduce drag. The plane had two main tires, and it would operate with one of them. If there was not a heavy load, it they could get rid of the drag from that flat tire. While they were cutting the tire off, there were a couple of helicopters that got shot down. One of them crashed in the middle of the runway. A couple of fighters were shot down in the area.

They decided, "We better call this off for the time being, anyway, because it's awfully hot for airplanes around there." They figured we'd lose a lot of planes. In the meantime, only one airplane, flown by a friend of mine, Major Ray Shelton, got in there. He picked up a load of people and got out. Then, they called it off.

There was another airplane called an O2, an observation airplane in the vicinity. It was shot down too, and crash-landed on the airfield. The pilot got out safely. Finally, the plane with the flat managed to cut it off the rim. They then took the ground liaison officer, which was an Air Force officer, with the Army unit, out of there. The guy flying the O2 took his place. The reason they did that was the air liaison officer was a brand-new guy, and he didn't know as much about the area as the O2 pilot, who knew the area like the back of his hand. So he stayed and they sent the air liaison officer out with a three combat control team. They were on the airfield, controlling the air traffic and the loading of the airplanes and to try to maintain order on the ground. Since evacuation was called off, they took those guys out. That airplane was en route to either Cam Rahn Bay or Saigon. They started the evacuation up again and that was about noon. That's when I got the recall to go back to Da Nang. I turned my airplane around. At the time, it was back down over Kham Duc and it was about three in the afternoon. The evacuation is going on hot and heavy at that time. Just before I arrived, however, the first airplane to land had all four engines shot out on final approach. In the meantime, before this happened, a young guy got out with a front-end loader and was scraping the helicopter that had crashed off the runway to open things up. It was usable but it was not completely open. The guy who'd had his engines shot out had had to crash land and ran off into a ditch. The next airplane took on a load of people being evacuated, but he took off in the wrong direction. He took off to the northeast up into the hills and he was not very far away when he got hit by anti-aircraft fire. He was shot down, crashed and burned and everybody on board was lost. The next airplane went in and picked up a load of people. It got out of there, but as I understand, he had over 1,000 holes in his airplane when he finally gave up counting.

*But he managed to get home?*

Yes, he did. All the other airplanes managed to get out, but all the other planes that went in there got heavily damaged. Somewhere around four in the afternoon and the weather had gotten rotten. There were thunderstorms throughout the area and also because of the burning of the camp, the visibility was low. I radioed in to the Airborne Command Post and they told me to circle at 9,000 feet, which I had been doing. When this one plane got off the ground, he called in and reported that he had picked up the last of the survivors. Shortly after that, the Airborne Command Post gave the fighters directions to go down and destroy the camp, to wipe it out. Another aircraft came on the air and said, "No, no, don't do that. I just let a three-man combat control team off." Those were the same men that they evacuated to either Cam Rahn Bay or Saigon, but when they had arrived there, they were told that the evacuation is back on, and to go back in and control the traffic. They went on another airplane, which landed right after the one that had taken the last survivors out. The combat control team got off, but there was no one there to load. After the pilot of their plane came across heavy attack on the ground, he took off and got out of there.

Now the camp is completely overrun by North Vietnamese except for the three American guys there. The North Vietnamese had set up automatic weapons positions alongside the runway, and all over the area, and finally after a long long silence, the Airborne Command Post sent a fighter down to take a look, to see if he could spot the Combat Control team, and he couldn't. They also sent down an observation plane to look, but he didn't see anything either. Finally, they asked another friend of mine, Al Jannotte, if he would go down and make an approach to see if he could draw this control team out of hiding. He said, "Okay." But the control team had been going from bunker to bunker and building to building, looking for survivors. They didn't find any. Then they made their way back to their control team area where they had their radio equipment. It had been destroyed so they had no communications whatsoever with the air.

Al made an approach to land, and I'm 9,000 feet over the airfield, and I'm watching him. He made a standard Air Force approach. I can see the tracers and the automatic weapons' fire coming out of the jungle and it scared the daylights out of me. He went on in and landed. He then rolled on down pretty close to where the helicopter was on the runway. He still hadn't seen the men, so he applied power and went around. Well, just as he broke ground and lifted off, he saw the guys running toward his airplane. He was low on fuel and had to go back to his home base, so he called out, said, "I spotted them, but I have to go back because of low fuel." Airborne Command Post called and said, "Is there anyone in the vicinity that can go in and pick them up?" Well, I had seen everything that was going on down there and I had figured I was the most logical person to do it.

*You were pretty scared, though?*

You bet. But I figured I had a better chance than anybody. So from 9,000 feet I did an assault approach, which is extremely steep. Very close to the airfield—a 270 degree, very, very steep spiral. I was able to land on the first couple of hundred feet of the runway. I didn't want to use reverse to stop quickly, so I just figured I'd stop with the brakes. The reason I didn't want to do that was that we had two auxiliary jet engines on the airplane to give us greater safety in taking off and, if I reversed the propellers, it would automatically shut the engines down. That's designed to keep from picking up rocks and sucking them and destroying the engine. What happens, the engine would shut down and the little doors on the front of the engines close to keep the stuff out.

Anyway, I was able to stop with the brakes. I stopped right opposite of where the three guys had taken cover in a ditch. As I turned toward them, they started running toward the airplane. Then I had to stop the airplane still so they could get on board. As I was looking back to see if these guys were on board, Major Jesse Campbell, observed a rocket that had just been fired at us from somewhere not too far away— a122 mm rocket. It came down on the runway and stopped about 23 feet from the nose gear and broke in half, but it didn't explode. Tracers were skipping off the pavement under the airplane. Fortunately, they didn't hit any of the guys running for the airplane.

As soon as they got on board, we pulled out and took off. No sooner had we got to where we were just breaking ground, the place where we had stopped just kind of erupted with mortar fire. They probably had just enough time to aim the tubes and drop a round in.We climbed out through rather heavy automatic weapons and anti-aircraft fire to about 3,500 feet. When it all stopped, I felt relatively safe at that point.

*That must have been about the most frightened you've been.*

Well, I don't know: it's pretty hard to measure one fright against another fright, and I've seen several of them in my life, but that was really shattering.

*What did those guys say to you?*

Well, I remember, what one of them said, he said, "I want to see you when you stand up because you must have a pair this big around." And he showed about the size of a basketball. I said, "No, that's not true." They were all drawn up on the inside.

*You did your share of monsoon flying, didn't you?*

Yes. Monsoon flying can be hard because most of the flying over there was what we would call VFR conditions. That means we didn't have adequate radio to do standard instrument flight rules flying. So we flew right next to the ground so we could see the ground, for example. I remember on several occasions flying under an overcast cast sky that was less than 100 feet above the ground and the visibility forward was zero. We could see out of the sides and keep track of where we were by looking to the side, but the windshield was so covered with rain, the wipers wouldn't wipe it off fast enough for you to be able to see. We just opened the side windows. We also opened the side windows for another reason: to let the rain out, as strange as it seems. The natural draft of that would draw any rain that came into the airplane, back out. And the airplane was not pressurized so it leaked like a sieve. Open the window let the rain go out.

*How long were you in Vietnam?*

Just a little less than a year. The normal tour was one year. However, there was a rule that I didn't know about. After you're nominated, you don't have to be approved, but if you're nominated and it gets past the first review level of command, you have to be taken out, or you're supposed to be. The nomination was made in the later part of May, and apparently, they forgot about it. I was due to come home around the 28th or end of August. At the end of July, I was getting kind of jumpy. I had been through all of the ten offensives and several other harrowing times there. I took some vacation and went over to Singapore for a week. When I got back, the guy that was substituting for me said, "Don't even unpack your bags. You're going home." It's like while I was gone, somebody discovered that rule and they wanted to get me out of there as soon as possible. They told me that, on the second of August, and I went back to home base down south on the third, packed on the fourth, and moved out on the fifth—a little over eleven months, almost a year.

*You also worked in Iran?*

I went over there after I retired from the military. This was to set up a maintenance-training program for the Iranian Air Force. I thought when I went over I would only have a training program with a couple of hundred guys, but as time went on, they bought more and more airplanes and I ended up with a training program for 700 or 800 men.

The Iranians had C-130's and they had F-5's and F-4's, and later on some Navy F-14 Tomcats. All of these airplanes were maintained by the U.S.A.F. system of maintenance. They wanted us to teach that.

Their Boeing 707 airplanes were kind of unique, in that they could do any one of three missions: air refueling, passenger hauling, or cargo. You could take the seats out and have a cargo deck, or put the seats back in. It would probably take a couple of hours to reconfigure.

*What did you think when you won the Medal of Honor?*

First of all, it's not a matter of winning something—it's a matter of probably making you custodian of something—a little bit of heritage. When you accept that thing, you accept a great deal of responsibility. You represent all the people who have done great things, made sacrifices and not been recognized at all.

*So that's a lot of responsibility, isn't it?*

Yeah. Also, you get yourself into a lot of things you would never do before. You have to perform work and go out and make speeches, which I don't think anybody really enjoys. But you feel you have to do it because of the honor that's been put on.

*You made an early flight across the Atlantic, didn't you?*

Yes, in 1950, I made two flights across the Atlantic on early missions of fighters. As a matter of fact, there were two flights before we took off.

*Anything unusual about that?*

I think everything about it was unusual. We had a guy whose engine blew up while he was over Canada, and he was able to glide back to an airstrip on the banks of the St. Lawrence River, where he bellied in and was picked up by an air rescue airplane. During take-off, they were trying to get off the ground and had a sort of near catastrophe.

*Can you explain further?*

He had made a belly landing on the airstrip and they sent in an air rescue to pick him up, an SA-16. The strip was short and soft and they figured they needed some extra power to get off the ground. They carried what we call JATO, jet-assist take-off, bottles. They attached those to the outside to give them extra thrust. When they started rolling down the airstrip to take off, they fired the JATO bottles. They rolled along in the soft dirt and mud and they figured, "This thing's not going to get off the ground before we get to the end of the runway, so we're going to abort on the runway." But you can't just shut those JATO bottles off, because once you light it, it's going to keep burning till it runs out. Then they reversed the propeller to try to stop it and put the brakes on. They finally came to a stop within a very few feet of the St. Lawrence River. This guy said it scared him worse than the belly-landing scared him.

The unusual thing about that was, that on the first flight across, we started out with 90 airplanes and we got 89 there. All arrived except the one that belly-landed. On the second flight over, we started out with 91, which was 90 plus the spare to make-up for the one we lost. And I think, as we took off for our final flight into Germany, that all 91 were rolling at the same time on the runway. It was probably the biggest gaggle of airplanes you've ever seen in your life. We started to release brakes within a few seconds of each other.

*What was the flight plan like?*

We went from Bergstrom Air Force Base in Texas to McGuire Air Force Base in New Jersey, up to Goose Bay Labrador, to Blue West I in Greenland, from there to Reykjavik, Iceland, to Manston, England, and from there to Germany.

*The sound of those planes must have been deafening.*

We couldn't hear it from inside, but we could sure feel it from the jet blast of the aircraft in front of us. I was right in the middle of it, and if anybody would have ever blown a tire, they'd have wiped out a dozen airplanes. We pulled out of that and up into formation, and went across that airfield in Germany. Ninety airplanes in close formation must have been a sight on the ground to see.

*Any strange planes?*

I flew what we called the pinball machines, which was an armor-plated P-63. They fired bullets at us made up of lead powder and Bakelite. They were supposed to shatter when they hit the airplane.

*Did they shatter?*

Mostly, yes. But often it did not shatter. Mainly you'd get a big shattering of the windshield. The windshield was pretty heavy though, one and one-half inch of tempered glass. On quite a few occasions, it got shattered. The bullets didn't go all the way through. It just shattered and you couldn't see through it.

*Ever in trouble because you couldn't see anything?*

No, but there were a couple of guys who got shot down. I never did. You had baffles to keep those bullets from hitting your radiator, but a couple of them got hit in the radiator and instead of shattering, went through, punctured it. They lost the coolant, the engine overheated and they had to belly in. They got out okay, though.

*It sounds a little scary.*

It wasn't until one of those bullets splattered right close to where you could hear it. You could feel them hit the airplane, but it wasn't that terrific until you got one right near the cockpit. It'd cause you to jump.

*Didn't you fly U-2's?*

I was one of the first Air Force U-2 pilots. I started training in November of 1956.

*What were those early planes like?*

The first one I saw was taking off was being flown by a test pilot. Remember this was 1956. The plane broke ground in a very short distance, and the nose came out at such a steep angle, I knew it was going to stall out. I started yelling, "Get the fire trucks—it's going to crash." Everybody in my group that had just seen it for the first time thought the same thing. He went out of sight and I was still screaming, "Get the fire trucks, He's gonna crash." But he never did. He kept right on going.

*Any other unusual things that happened there?*

I was heavily involved in the initial planning of the missions over Cuba. That was in 1959 and Cuba was getting restless—things were going on there. I was in the Director of Plans in the Reconnaissance Wing the time. About three of us were sitting around, thinking of what we should be doing and what came up as one of the things we should do was go down and take a look at Cuba, and see what was going on there. I drew up some plans for Reconnaissance of Cuba and sent them forward, but they were rejected. As time went on, it became more and more obvious to us that we should be doing something about Cuba. So I sent the plans forward again and they were rejected again. In 1960, after Gary Powers was shot down in his U-2, I was transferred to Headquarters, Strategic Air Command, to be Operations Staff Officer to the operation of the Air Force U-2's.

One day in August or September, I had a phone call from a guy in the joint reconnaissance center in Washington. He asked me if I remembered those missions. I said, "Yeah, I sure do." We had a code name for them and he referred to the code name. And he said, "Can you still do them?" I said, "Yes." He said, "Okay, I'll send you a message." And he sent me a message, outlining a set of coordinates in Cuba that they wanted covered—and that was all. So I drew up the mission. And, in addition, I drew up the same two missions that we had planned earlier, in 1959, and included those in the plans. I took them to Washington myself, and I did the briefing at the Joint Reconnaissance Center and they took the plans and said, "Okay," and sent me back to Omaha. In October, I was downstairs at SAC and my boss came in and told me they wanted to execute the Cuba missions. I went out to California where they had special airplanes. In the meantime, we alerted the aircrews and moved the people necessary to do the mission. Anyway, we got in late on Friday night and started planning the first mission early the next morning. After planning, we launched Richard Heyser early Sunday morning on October 14, 1962. Later that afternoon, we learned he had safely landed at a base down in Florida. We got the word that he had recovered and that everything had worked okay. Later, we got a phone call from the Chief of Staff at SAC saying, "Get out to Florida and get prepared to launch the other two missions." He sent a jet tanker for us and it took us to Florida. We spend all the night, Sunday, planning the next two missions in more detail, the next morning, which was Monday, 13th of October; Major Heyser was again the first airplane to get off the ground. Fifteen- minutes later, Major Rudolph Anderson took off for the third of these series of missions.

We had the day off the following day and the planes got back that afternoon. We sent the material to Washington, D.C., and then we had the day off to get some sleep. Some of us hadn't slept for about three or four days. Following that day off, they started flying hot and heavy, because the three missions were 100 percent successful. The first mission located the first offensive missile site in Cuba; the next two missions located every other operational offensive site that had been built or was under construction in Cuba.

We were working frantically there for several days. I think it was on the morning of October 18 that I got a call from my boss back in Omaha that he wanted me back there right away because he needed me. That was all he would tell me. He says, "I'm sending an airplane for you." He did. While I was getting ready to leave, I ran into Major Anderson, who was the pilot who flew the third mission. He had been flying regularly the rest of the week. As I was getting ready to leave, he said, "You know, Joe, I'm not going to come back from these missions." I said, "Hey, that's no attitude to take. You take that attitude, you're sure as hell not going to come back." He said, "No, no, I'm serious. I'm dead serious. I've got this premonition. I know it and my wife knows it." And I said something to the effect that I would say something to the wing commander and get him taken off these missions. He said, "No, don't do that. I'll tell him that there's not a word of truth to that, that you're mad at me and trying to get even with me. Not only that, if I don't fly these missions, I'll feel like I'm a traitor to my country. The Air Force has paid me for fourteen years, and I think they've done that specifically so I could fly these missions. If I don't fly them, I'll be worse than a traitor to the Air Force and to this country, and I can't have that happen."

*That's a heavy burden.*

Yes, the pressure at that time. I didn't have time to do anything about it and I don't know if I'd have done anything about it anyway.

That was something that I've wondered about since them. Anyway, I went back to Omaha that day. It went on to around ten days later. That was the day when President Kennedy laid his speech to the Russians, outlining his points.

What SAC was doing and the reason they wanted me back there, was they were activating the battle staff, which means they were preparing for war. And they wanted the battle staff activated before Kennedy made his speech that afternoon. I was a member of that staff.

We had a detachment of U-2's up in Alaska at the same time. They were doing air sampling missions up over the North Pole. One of the guys had a compass malfunction while he was on the flight. When you get to the North Pole, it's not one direction to go. You've got 360 degrees, 360 South's that you can take. This fellow took the wrong direction and got lost. We finally found him and, through an air rescue airplane, got him headed back toward Alaska. He was almost out of gas and was gliding back. He was not completely out of gas, but he had very little left. Finally, he was letting down and was trying to get into Nome, Alaska. At low altitude, he tried to start his engine, but it wouldn't start. He broke out of the clouds, just precisely so he could look over things and there was the runway at Nome. He did a dead stick landing at Nome. He hadn't been on the ground more than a few minutes, when one of the guys came in and said, "Hey, have they found the airplane yet?"

I said, "Yeah, they found it on the ground in Nome, Alaska. He just landed dead stick." He said, "No, no, I'm not talking about the one up in Alaska, I'm talking about the one over in Cuba." I said, "That was Major Anderson, wasn't it?" He said, "Yeah, how did you know?" I said, "He told me ten days ago the wasn't gonna return from this mission."

*What's flight meant to you?*

It's probably been most of my life. The only thing that was more important was my family.

# An Encouraging Voice

## Sheldon B. Benscoter

Sheldon Benscoter is a vigorous man, still surprisingly vital in his senior years. His voice is strong, even powerful, though there is warmth and quiet confidence to him. Born in East Maunch Chunk (now Jim Thorpe) in 1921, he served as class president, graduating from high school in 1938. He decided he should know shorthand and typing, so studied it. Next, he got a job making handbags, but that was short-lived when the company shut down. They couldn't get needed materials from Egypt. The National Youth Administration offered potential opportunity for the future, so he engaged with it. At night, he and his colleagues would travel thirty miles down to Bethlehem for tech school, which meant four hours of classes and four hours of shop. Guaranteed a job at Bethlehem Steel Corporation upon completion of an eight-week course, he was able to go to work as a machine hand, when the government shortened requirements after seven weeks.

Sheldon B. Benscoter ended up making the rounds when he joined the military. First, he was sent to Miami Beach for Basic Training. Then he went to Cleveland, where he took some college classes and got ten hours in a Piper Cub. Next was Nashville, where he classified for pilot training and on to Montgomery, Alabama.

There, he met disappointment: he washed out of pilot training after 65 hours. Next stop: Valdosta, Georgia. At Tyndall Field, he learned gunnery and, after that, to Victorville, California, for bombardier training. That led to six-weeks tacked on for instruction in celestial navigation. After a furlough, Sheldon returned to California, to March Field for bombardier and navigation training. He thought he'd end up in medium bombers, but he ended up in the big, four-engine, B-24 Liberators, assigned to the 98$^{th}$ Bomb Group, in Italy.

He was recipient of Air Medal with two Oak Leaf Clusters. European Theater of Operations Medal with five battle stars, Prisoner of War Medal, Good Conduct Medal, American Campaign Medal and World War Two Medal.

After returning from the military, he graduated from Muhlenberg College. Active in civic life and love of community, Sheldon worked as a reporter, copy reader, assistant wire editor and National Editor over a period of thirty-five years for the *Call-Chronicle* newspapers. He died in 2005 with his loving wife, Marilyn, by his side.

*How'd you end up in the military?*

That program resulted in a job in a defense plant at Bethlehem Steel, and all my friends were being drafted and joining. I guess I was listening to too much patriotic music, so I decided to join the Air Cadets. I took the exam and flunked it. The reasons were that I was under-weight and needed an operation on my nose. But I impressed the recruiting sergeant when I went and had an operation for my deviated septum on my own. He was impressed that I paid seventy-five dollars for it myself. As a result, he waived my weight requirement, telling me to eat a lot of bananas.

*You washed out as a pilot. Was that traumatic?*

What happened was that I had ear trouble and some others and I were grounded for two weeks. I had caught a head cold and my ears wouldn't pop. The military doctors tried to open them, but they couldn't. We were doing aerobatics at the time. That may have done it. We were going to be washed down to another class. Then someone decided we were to graduate with our class, so we had to catch up. But, when I flew on a check ride, I was so nervous I didn't do well. I got sick and the instructor said, "Roll down the window." I vomited. The instructor informed me that I hadn't made the cut. I still had an hour of flight time scheduled, so I said, "Can I go up for an hour?" He said, "Yes, if you want to kill yourself" (Laughs).

*How many missions did you fly?*

We had to fly 35 sorties or 50 missions. Flying into Germany was heavily defended flying, so they would give us credit for two missions, one for each way, -- one for getting there, one for getting back. I had 28 sorties, or 42 missions, which means I had fourteen that counted as two.

*Most dramatic mission?*

One mission to Vienna, Austria took us over a lot of targets. We were following the railroad into Vienna. We started our bomb run, and I looked up and the plane in front of us was gone. I heard a thump between my legs. When we got back to the base, I checked and there was a hole the size of baseball where the shift was. If it had been a foot or two higher, I wouldn't be here today.

*Did it make you get religion?*

We had one guy who had a bunch of flak come up his side and knock his hat off. He started going to church.

*Another terrifying flight?*

We were bombing a bridge in northern Italy. The heavies had a try at it and came back, and then they sent out the medium bombers. When they couldn't destroy it, they sent us up there. We reached the initial point and I could see the bomb group going in. One of the hardest things is watching the planes ahead of you and seeing what they were going into. All of a sudden, the main plane got a direct hit. It was just a flash and the wings fell off. Nobody got out alive. That night we did a lot of commiserating at the bar, where plenty of promotions took place. The big shots had to go on those flights, so nobody would end up sitting back at base, reading about it.

One time I was flying with another crew. I had replaced the bombardier. Something was wrong with him. This was on a flight to Vienna and a big oil refinery. They said five or six hundred anti-aircraft guns were waiting. And as we approached the drop point, I couldn't get the bomb-bay door open. I asked the pilot what to do, and he said, "We are not going to bring the damn things home." I said, "Drop them through the regular doors?" I did that, and as I did, the wind tore the door off. The temperature was 54 degrees below zero and everybody got frost-bite on their faces, around where their masks were.

*Did you get an acid stomach on flights?*

No, I didn't. I had a ritual. I only had a cup of coffee. Someone told me if you got hit in the stomach, you stood a better chance of survival if it was empty (Laughs).

*Any real characters with whom you flew?*

We had a pilot who had this little goatee and it was a nuisance when he was always scratching it. He wouldn't cut it off for fear he'd lose his luck. It was like a good luck charm. One flight he was hit. He tried to fly with one hand. He was thrashing around in the seat. When the flight arrived back in Italy, the survivors had to bail out over the airfield there. They said the co-pilot's hair had turned gray. That was the last flight he took over there. They sent him back to the states. His nerves were shot. He had flown four hours with this guy thrashing around, dying beside him.

*You weren't on that flight, were you?*

No, but on our last flight, we inherited the last two guys on that crew. They thought our mission was a milk-run, but one of them was killed; another was captured with us. That old crew was considered a hard-luck crew.

*What did you think when they joined you? Was it scary?*

We didn't know who they were when they first joined us. They just came off a plane and didn't introduce themselves. You didn't know the other crews so well. It wasn't until we got shot down did we know.

Our ground crew had a little pet dog. It was gentle as a lamb. It would play with us before we would take off. That day, we played with the dog, and as the intelligence officer was pulling up to take all our personal belongings and give us escape kits, the dog ran in front of him and was hit. The crew chief had to borrow my gun to put it out of its misery. That cast a shadow on our last mission. We were all down in the dumps for loss of that little dog.

*You didn't say, "No flight today, please?"*

No, the mission took us to a railroad pass about eighteen miles from Bremen. As we started down, we were hit and it knocked out two of our engines. I pulled out the nose gunner and he and the navigator headed out through the tunnel. I followed and got to the bomb-bay. By now, the navigator had opened up the flight deck and he was yelling, "Jump." I pushed the nose gunner out the catwalk. He jumped and I jumped. I started to count to ten before pulling the ripcord on the parachute, but when I saw the mountains, I pulled it at six. Nothing happened. I had my back to the ground. I could finally see what they called a "pilot chute." I pulled it out and it caught the wind and opened. As soon as it opened, I started to sway back and forth like a pendulum. When I got to one side, the parachute panels on the other side would collapse and I'd drop. I could see trees and mountains, and soon I landed. When on the ground, I did a barrel roll and started on my feet. I guess I was fifty feet from a little farmhouse, and a man in black Sunday suit—it was Sunday—came walking toward me. He seemed to be looking right through me. I looked back and there were four German soldiers, so I put my hands behind my head. They took my .45. They let me get my shoes off the parachute harness and we marched off. A kid came running along beside us, asking for my shoes. The German drove us up the mountain, and he had the horn blowing all the way. We were met by an officer in black, and I saluted him, as is military custom. Then I noticed he had a skull-and-crossbones, and I thought, "Shit, it's the SS." They took me in a room with a long table and told me to undress. A woman in uniform came in. She wore a monocle. I had gotten down to my shorts. I declined to remove them while she was there, and finally she left. The officer gave me this look when she left, and I knew what was coming next. They were going to look up into my posterior to see if I was hiding anything. They found nothing and gave me some coveralls. All over the walls were old spears that looked like the Middle Ages. A soldier said to me, "For you, the war is over." I could see one of our planes as it flashed by in the sky, going home. I felt lonely as hell. A German soldier yelled to get away from the window. They took me into town, where I was interrogated. They told me, if I didn't try to escape, I could go to the downed plane and identify any bodies. We were told not to reveal information, so I declined. They took me to another room where Kinsley, from our plane, was sitting. At first, we pretended not to know each other. Then they brought in another of the team. He had bailed out of over India and didn't have his parachute tight enough, which had done damage to his private parts. In fact, on our flight, he had frozen on catwalk and the co-pilot had to push him off. The co-pilot, Larry, tried to get away but the Germans fired at him and he was captured. We didn't know what happened to the pilot and navigator, Patchen and Paulsen. There for a couple of hours, we saw some scorched chutes and figured they had burned. We had escape kits in the pockets of our flight pants and, when they started to search me, I yelled, "They're searching me again." I was hoping the guys heard me. They did and hid one in a potbelly stove. The Germans tried to take my ring, but I claimed I couldn't get it off, so they let me keep it. I had a locket which held the 91st Psalm and a lock of my mother's hair. "A thousand shall fall with the right hand . . ." it read. They let me keep it. I still have that locket to this day, in a little drawer here. I also have a small, hairy monkey I got at Catalina Island. I carried it for good luck. The Germans let me keep it. I still have it, after all these years.

*What are you thinking about at this point?*

I'm thinking I'm lucky to be alive. I discovered the guys had heard my warning and hid the escape kits in the stove. We took the maps from it and with the sewing stuff sewed the maps inside our jackets. An hour later, they brought in gunners from another crew. They were covered with pink hydraulic fluid. We sat on the bed and could see something burning. We never learned what. The next day, they brought in a navigator who had almost landed on top of a flak factory. Then we found out that our pilot was alive and that he had ended up in a hospital in Murano. Then we were in, I think, Bolzeno. People across from us were Yugoslavians, whom they held as slave labor. We ran into two more prisoners, one of which was a pilot on a B-24. He remembered only standing behind the windscreen, trying to pull back on the stick to pull out of a dive, when the plane blew up. He was the only one to get out alive. He tried to get over the mountains, but then in a couple of days, he gave up. It was very depressing. All you could see were mountains. We had five German soldiers with us, and we marched out. There was an air raid and they put us in a shelter, which was great, because they'd never done that before. One guard in charge spoke good English and liked Gershwin's music. He asked us to sing him some Gershwin songs, so we did. In exchange, the Germans sang, "Lili Marleen." "Our two shadows//looked like one." Next, they took us to a railroad line, and the rails were twisted like hairpins. We continued on and got into Austria. A young girl gave us water. We marched and marched and finally came to a convent and the nuns came down and started to feed us. We finally marched to Innsbruck, where a big air raid had taken place. Dust still hung in the air. The Nazis lined us by a wall and kept everyone away from us. Next morning we started out through all that rubble. You could smell the stench of death. We got on a train and I remember going through those rolling hills that were like Pennsylvania. I remember seeing a deer. When we ended up on the outskirts of Munich, there was strafing and rubble was all around. GI's told us the prisoner camp was ten miles away. It turned dark and we came to a brick building with a big chandelier and big banquet hall, filled with officers and their wives dressed up in gowns. We found out they were observing Hitler's birthday. They took us to the kitchen and we had a little broth and then to a railway station. The German in charge loved photography and he had two cameras. He showed us photographs, including one he took during the Nuremberg Rallies. It included Hitler, Goering, Himmler, all the big important ones, and included a guy named Ernst Rohm of the S.A. (Storm Troopers), who was executed by Hitler. One of our guards was in the photo, flanking all those people. A little old lady gave us a cookie. We finally reached a barn and, when the door slammed behind us, it was scary. But then the door opened on the other side. Before us sat the prison camp, Stalag VII-A in Moosburg. It was drizzling at the time, and a large line of GI's waited for soup. Rain kept filling the soup bucket as fast as it had started to empty. We ended up in barracks where we heard they housed Russians. I woke up that night with fleas. They gave us showers and we put on the flea-infested clothes. The Germans wanted to know my job on the plane. I didn't tell them I was a bombardier, because I would have been on their shit list. I roomed with a P-47 pilot who thought he was landing on an Allied field, but it was

German. He has cracked his head and had a terrible concussion and went in and out of consciousness. The Germans took him out so I had the room to myself. The Red Cross parcel that came contained, among other things, three cigarettes. I traded the guard those cigarettes for a pocket watch, a knife and a hard-boiled egg. I still have the knife and watch. The Germans wanted us to pose for pictures with them. As the Germans retreated, they took POWS with them. We had to lay on the ground wet with rain. They'd taken all the wooden seats off the toilets to burn as firewood. Paulsen was handy and he built a stove, and we went looking for firewood. Somebody was listening and the colonel said there was a fight. The SS wanted to shoot up the camp. Then we saw the guards running away. It was General Patton had arrived. A short GI was sewing a flag to his jacket. He had been captured in Africa and said he had waited to do this for three years. He was crying, and I cried, too. It was Sunday, April 25, 1945.

*How long were you a POW?*

About a month.

*How much did you weigh?*

I don't know. I know I weighed 145 pounds when flying. In September, 1945, I went to Camp Dixie for release. I was 130 pounds, after eating heartily for two months.

*How about the arrival home?*

We took a troop ship home. And there was the Statue of Liberty. I took a train home and my mother, father and sister spent half the night talking. They hadn't heard from me and, at one point, thought I was dead. Our congressman said someone told him they had seen our plane go down and that no parachutes were seen.

*How long had they assumed that?*

I got back June fifteenth. I think my aunt had discovered a list of POWS in Philadelphia. They believed that probably for two months.

*What is your proudest moment?*

My proudest moment came a year later. I was working in the Pokonos, and my pilot said that all the guys he talked to said, "Thank God for 'Scooter'." "Scooter" was my nickname. They said that I kept them all alive. We had five guys on the crew and me, along with four other men on the trip from Italy.

*What did you do?*

I reminded them that eventually we'd be safe, we'd be liberated. I'd tell them jokes. We'd sing-along. When Franklin D. Roosevelt died, I said, "Don't worry, someone else will take over Roosevelt's place." We learned about him when the Germans showed us the newspaper that read: "Roosevelt Mort."

*Were you conscious of your leadership role?*

I was a senior. I don't know. I often wonder if I could do it again today. A couple of times we had a green guy flying and he's praying all the way down the bomb run. Heard many "Hail, Mary, Full of Grace, The Lord is With Thee . . . pray for us sinners now in the hour of our death."

# Of WWII and Korea

## Wes Hazlett

Wes Hazlett spent 28 years in the Marine Corps and flew in both WWII and Korea. In 1941 he graduated from Muscatine Junior College, where he took a flight class in CPT program. In the fall, 1941, he intended to enlist in the Army Air Corps, but was turned down. Not to be deterred, he went to see the Navy Examining Board when it came to town. Wes had the necessary two years of college and had completed some flight training, so they accepted him—without even so much as an x-ray. He was sent to St. Louis for pre-flight, then on to Corpus Christi and Brownsville for fighter school. He spent four months and two days in flight school and finished with 196 hours of flying time. Then he went to Opa-Locka for the final transition to fighters as a gold-bar cadet. At the time he was a second lieutenant and chose to join the Marine Corps, which was optional in those days. Wes wanted to be in the Marine Corps because they had primarily fighters—no big boats or anything like that—and the Marines selected him. In Miami he transitioned to the Brewster Buffalo (F2-A), an airplane used in the Battle of Midway. The airplane offered considerable performance, but the plane did not have self-sealing tanks so U.S. fliers lost nearly all of them on the first day at Midway. It did not take long for Wes to encounter his own problems with the aircraft as he explains:

On the Brewster the engine manufacturers beefed up the engine to 1200 HP from the intended 900 HP. The process overtaxed the bearings in the plane, so the engines would immediately heat up and seize. We lost a couple of planes from that. I was flying out about ten miles from Miami when it became apparent the engine was going to keep slowing down. I tried to crowd the throttle but could not get any rpm's out of her. I thought I could make the field, so I put the gear down. I landed in the sand outside the field, but the plane did not turn over. I guess I was going so slow it did not have enough energy to go end over end.

Local mechanics installed a wire on the Brewster as a sort of emergency procedure. When anything went wrong, the pilot was to jerk on the wire. I never did figure out to what the wire was connected. I think it regulated the mixture control, but they never told us. We were naïve citizens about that time. The military just took us and shot us out into the wild blue.

The Brewster was the first airplane I ever blacked out in. You would take off and by the end of the runway—you might have 3,000 feet or 4,000 feet—you had that plane up to speed and when you pulled back on the stick, you would feel it climb. Another unusual aspect is that it would fly sideways. When you trim most airplanes, the ball will stay in the center. With this plane you would supposedly get it trimmed up, kick the rudder and it would stay there. You would kick the other rudder and it would get over on the other side and stay there. The "flying peanut" had little directional stability.

The airport made a wild gunnery platform. It was almost impossible to keep the ball in the center for gunnery. I remember the first time I blacked out. In those days, the military used to teach you to make overhead deflection runs. Often, in order to preserve your good health, you would have to make it a high deflection shot from an overhead position, say about 12,000 feet. In training you would go down past the target banner at 8,000 feet—straight down—and then make the pullout. The first day I did this I pulled back on the stick and that was the last thing I remembered. Everything was turning blue and gray and finally nice and blue again. The blue happened to be the sky. By then, I was pulling past the target. I had been blacked out for about ten seconds. It taught me a lesson that you have to do a little grunting and groaning to keep the blood up in your upper body. Of course, in those days we did not have a G-suit. Down in fighter school we used to fly the SNJ and the AT-6 where you would not pull the G's for so long. When I got into the Corsairs, I needed to grunt a little better. We used to have competitions to see who could stay on whose tail in a wind-up turn. A wind-up turn is where you get the airplane spiraling down and keep tightening up on the G's. You can actually sustain the G's for a long period of time and feel yourself dim out. You felt your vision begin to tunnel and get narrower and narrower. Finally, you could not see ahead. You had better back off a little bit then, because right after then, you are not going to be looking at anything.

*In what ways were you naïve in those days?*

One was that I never learned to swim. We went through flight school in 196 hours so it was pretty concentrated. You would fly once or twice a day. There was a swimming class, but I had come out of the Midwest and could dog paddle about 100 yards, but that was it. Here I was joining the Marine Corps where I might be a long way from the shoreline, but I never did learn to swim to any degree. Finally, when I crashed into the Pacific one time, I found out why you had to know how to swim. After the fighter training in Miami, we were sent to the West Coast where we were carrier qualified on one of the little carriers of the day. The carriers had a deck about 60 feet wide on them. Those carrier qualifications were in the SNJ. From there I was sent overseas. I joined a squadron in Hawaii—VMF-222, which had just come out of Santa Barbara, along with its sister outfit 215. We were based on the western side of Hawaii, west of Oahu, at Eva. There we played around a little bit with the Army, which was flying out of Hickam Field. That was where I had my first Corsair quit on me.

The early Corsair did not have adequate pressurization of the distributors and as a result, at high altitude, the ignition system could fail.

You would get an arcing across the distributor, which would burn a trail across the inside of the distributor caps. Then you would get no more spark to the engine. The Navy had their airplanes launched someplace else, so we Marines were sent out to cover the carriers. I reached the location and was cruising around about 25,000 feet when suddenly the engine quit. Unfortunately, we had taken the hooks off our airplanes. We had removed all excess weight from our airplanes to make them perform better. The Navy was not yet using Corsairs, which had dangerous landing characteristics. Anyway, I kept trying to start the engine up, but nothing would happen. The engine would just windmill. It takes awhile to come down from 25,000', and when I got down to 3,000' the engine coughed and started running again. I was just about ready to ditch the airplane, because I figured I was not going to get the plane onto any flight deck with a hook. After that, the mechanics started pressurizing the distributors a little better, and we never had any trouble like that. That was in 1942.

*Did you have any rough landings in the Corsair?*

No, I never had a problem. I never scratched an airplane until later on.

*Did you see a lot of accidents?*

I saw a guy one day do a snap roll in the final approach. As you know, a snap roll is just a stall and a spin in a horizontal direction. Anyway, this guy was coming in and as he pulled in tight to make his final approach at 200 feet, he did a complete snap with the airplane and landed on his wheels. It sheared the runway and tore his wheels off, but he landed right side up.

*Did you lose a lot of friends in the war?*

One of my good friends I lost was Jess Leech. He went down at Rabaul. The pitiful part of it was that those guys would be in the water and we would try to save them. Frequently the pilots would be under fire from the shore, and unless we got "Dumbo," the float plane, out there quickly, the enemy was going to come out and pick them up. Of course, very few people ever came back from Rabaul. Each was in our flight when he went down about a half-mile off shore. There were about two hours of daylight left before we had to land. "Dumbo" landed, but they were under fire from shore batteries, and could not wait around to get him in the boat. So Jesse just waved and that was the last we saw of him.

*You saw him wave?*

Oh, yes.

*He waved goodbye—that must have been a shock. How did you respond?*

It's enough to bring tears to your eyes, isn't it? But you can get pretty hardened. You become close to only a few people. As you lost pilots and the replacements come in, the relationships are different.

*Describe your friend as he waved goodbye. This is like an admission he would die.*

No, I don't think that was the case. I think he was letting us know he was all right, and that he knew we were trying to help him. It wasn't that he knew he was dead, although in the early part of the war, not many pilots got back from Rabaul. Neither side was burdened with prisoners. When we first went out into the South Pacific, we went into Guadalcanal. Then, as the war progressed, we moved our airfields up the chain—to the Solomon Islands, up through Munda to Vella Lavella and Bougainville. When we got to Bougainville, I visited a Marine Corps camp, which was a sign on four posts with barbed wire around—a Japanese prison camp. The confines there were about a yard square. Our pilots would end up nailed down to a stump, disemboweled and pushed over backwards. I never saw what happened to Japanese prisoners.

*Did you see guys like that?*

I did not see guys like that. We did have a pilot whose face had been pulled back—his scalp pulled down over his face.

*Where was that?*

That was in the Philippines. That was a Marine captain that was shot down. He was paraded through the streets of Cebu. We went in there after the place was secure—just barely overrun, and the locals related what had happened to our pilot.

*Does it make you sick?*

It doesn't make you sick. It makes you angry. You realize how brutal war is. It is something you never want your own kids to go through. The attitude that I went to war with involved an inner faith that whatever will be, will be, and that I would try to do what I could, do my best to get through it. But you have got to believe that you will luck through it. I left the states several times and went overseas for 26 months during WWII, and every time I went out, I would look at the Golden Gate Bridge or something like that and wonder if I would ever come back.

*Did things ever emotionally catch up with you?*

No, but I was always scared. When we would go out in the morning, especially in the Solomon Islands, we knew we were going into heavy enemy territory. I could not eat breakfast a lot of times; however, when I came back, I would go right to sleep. We always had a doctor with hospital brandy. Strangely enough, I could relax completely. For me it was complete relaxation when I got back, even though I was living in a mud hole.

*Some of the other guys must have stayed up half the night.*

Some of the guys in our sister squadron—214—Pappy Boyington's squadron—would be carousing all night. A lot of our guys would be up carousing. But I never did drink much. I couldn't drink with the big boys. The personalities of your squadron mates were all a little different, but they were always respectful to me. We all had different ways to relax.

When I joined this squadron in Hawaii, I was the new boy. The first thing that went wrong occurred after we went to Eva. The military expected to ferry us to Midway, where they were expecting another attack. First, I was supposed to fly an airplane over to Ford Island at four a.m. and exchange it for another. Then the next day we were going to go out to French Frigate Shoals and then to Midway. I came in from Eva, about 20 miles west, heading to Ford Island, and I looked down to see the airfield lights. The controllers called in a traffic pattern, so I started to land. Ford Island is right next to Hickam Field. Though the lights were on down there, I did not know which field was which. Finally, I landed the damned airplane and as I pulled up to the tower there, I said, "Hey, where do we take these Navy airplanes? We're going to exchange." The guy said, "You're on the wrong airfield." Here I was on Hickam field! By the time I got to the right base, my name was established. They said, "Okay, from now on, you're "Hick" Hazlett." So they had a brand-new boy with a brand-new name.

At Midway they assigned me to a guy by the name of Willy, the division leader. You have a division leader and a section leader in a four-plane element. I was Tail End Charlie. Tail End Charlie flies behind everybody else and is the first bait in case somebody comes up behind to shoot and is supposed to follow and protect his leaders. Anyway, Willy was the hottest pilot in the squadron. He would bet you 100 dollars that you could not stay on his tail for ten seconds. He used to take the airplanes up over Midway and pull up into a hammerhead stall—that is, pull straight up, kick her over and stall her out and go straight down. By the time the pilots below him would start up, Willie was going the other way. Nobody could stay on his tail more than ten seconds. One of our leading aces, a guy named Sapp, took the bet too, but Willy took his money. Since I was Tail End Charlie and one of the guys who could stay behind him most of the time, they called me "Wee-Willy," so I got a new nickname. In the book *Hell and the Heavens*, I am "Wee-Willy." The trouble was Willy was about five-foot-six and I was about six feet.

*That must have been a relief to get the nickname" Wee-Willy."*

I got rid of the first nickname after about two months. It was a great relief. When we went down in the South Pacific, we had a skipper who was good in terms of providing for his squadron. He got us a laundry and other things, which the other squadrons did not have. But I do not think he was a very good squadron leader in combat terms. Anyway, this skipper had the idea that none of us junior bird men were adequate for the big stuff, so he would not let us fly the strike flights. Instead, we flew the dawn and dusk patrols. Once he sent a few of us junior bird men over to attack Kahili Air Field on Bougainville Island. At the time the skipper and his squadron had gotten into a scrape and started home. One plane went down in the ocean, but three of them reached home. Two of those three were running out of gas and got in a traffic pattern going in opposite directions. The two pilots were anxious because they were running on fumes. As the two came in to land from opposite directions, one of them had to take a wave-off. When he did, he ran out of gas and fell in the ocean at the edge of the runway. That day the military lost three airplanes out of four. Thereafter, we all got to fly everywhere.

Normally, we would fly for six weeks. It was pretty rough. We'd be flying hard every day and staying up all night dodging bombs and "Washington Machine Charlie" in air-raids. Then we'd get six weeks back down at Espirita Santos for further training. Another squadron would come in to take our place. The skipper lasted only the first six weeks, when they decided to make a change. About that time Pappy Boyington was coming in and the rumor had it that he was going to get the outfit—the 222. Instead, they gave him the 214, and he got a bunch of renegades of his own in there, but we were based with the 214 up on Vella Lavella. Boyington's guys would do a lot of carousing around, but Boyington was a good combat leader.

*His reputation is legendary.*

He'd drink all night and the next day he would get up and fly. There is only one way to shoot down an airplane and that is to get in behind it and start shooting. He'd do it. Of course, being squadron leader, you were the first guy there. By the time Tail End Charlie got there, the enemy was mad. It was like stirring up a bunch of hornets. And the Zeroes usually could outturn us. If you didn't surprise them and get out of there, then you had better get out, because it would not pay to stay and mix it up with the Japanese.

In fact, I think that's what happened to Willy Moore. Willy and I went out on dawn patrol. The radar picked up a flight of enemy coming in towards Vella Lavella. At that particular time, we were extended out in a scouting line, which meant that we were spread out laterally with about a quarter of a mile between each of us. I was sitting at the right end of the line. When we tally-hoed, it just so happened that the enemy—a bunch of bombers and fighters—was on my end of the line. Willy was in the center and the other guys were on the end. By the time we saw them they were up with us and off to my side under a layer of broken scud. We turned our Corsairs to attack. I was the first guy there and started to tail in on some Bettys bombers. About this time the Zeroes were all over me, coming in from both sides. In fact, all I could see were radial engines pointing in on me. I pulled up into a cloud layer above me and made a 360-degree turn and came back out of the cloud. By then the Japanese aircraft were ahead of me, so I started shooting at them. I was the only guy left. Willy was gone, as well as Moore, and Carroll, the other guy in the flight, who had his canopy shot off and headed home.

*They both went home?*

No, Willy Moore must have stayed, but he never made it. The only thing I can figure is that Willy thought that he could outfight anybody, and he got in there and tried to outturn the Zeroes on that first turn and somebody got him. I did not see him go down, but he never came home. Moore and Carroll did come home. But that was it: Willy's first flight was his last flight.

*How many planes do you think there were?*

Thirty airplanes. About six bombers and the rest were Zeroes.

*How close were you? Could you see the enemy pilots?*

You can see pilots and you can see the gunfire spitting at you. All I remember really is how often the big old radial engines came pointing at you. You wanted to be pointing at them.

*Describe your escape there. Were they still after you?*

Oh no. When I made my 360-degree turn into the cloud, the flight of Japanese moved on. Of course, I was sitting behind them. As I came back out of the cloud, their stragglers were catching back up with their formation. I saw a straggler, a Tony, sitting right in front of me. I started shooting at him and pieces started falling off his plane. His airplane just languished there and rolled around a little, and then it rolled in a split-S and went right on down. I followed him down.

If a Zero saw you before you got there, they were going to be turned around and headed right at you. Then you would be heading right at them. I had been in a number of those escapades, when I was "Tail-end Charlie." One time I talked to my section leader, Jules Koetsch. We decided that the first guy that breaks during a head-on attack was going to get shot. In other words, as long as you were heading at one another and shooting, you had that big engine up ahead of you, as well as bullet-proof glass, but the minute you tried to turn and break off, you exposed your belly and were liable to get shot. The idea was to play chicken, to run in closer and closer and when the guy would chicken out, he would break off. I told Jules, "Okay, I've got things all figured out. The way the airplane turns the easiest is just to pull back on the stick. Even if you want to turn right or left, you will roll the airplane and pull back on the stick, because that way the forces build up on the wings" I said, "Okay, Jules this is what I am going to do. Next time I see one of these guys coming at me, I just going to drive it right on in there and then at the very last second, rather than pulling back on the stick, I'm going to push forward on the stick and go down under him."

We went back to Esperito Santos on a six-week break and were playing around there. One time we decided to take 12 airplanes up in the air and dogfight one another. So we got to doing that. Dogfights invariably end up with every man for himself—even though we might start off as four-plane elements. I looked across at the other side of this melee and here came a Corsair. I decided this was the time to try my experiment. I went flying at this Corsair. He was flying at me. At the last second, I pushed forward on the stick, and the other guy did the same thing. It turned out the other guy was Jules. The only thing that kept us from colliding was that our wings were not quite parallel with one another. But we did fly by one another at 500 miles an hour—closing speed there was about 250-300 mph each. As we did I looked at him right in the eye. It was canopy to canopy.

*You should not have given him the idea.*

I gave the wrong idea to the wrong guy. On anybody else my experiment would have worked, but with him, it just scared both of us to death.

*What did you say afterward?*

Oh boy! He and I were pretty frightened. That was about as close as we wanted to come in that game.

*Did anything unusual happen in training?*

I lost an engine. I had a problem down in Turtle Bay where we flew for this six-week period. Turtle Bay was this little 3,000 foot runway on this plantation, which had coconut trees on the sides of the runway. I was "Tail-end Charlie," so I followed after everyone took off. The idea on the join-up is to get there just as fast as you can, then cut back on the throttle and slide into position. I was booming up through there with my throttle wide open in the process of catching everybody, when I went to pull back on the throttle and nothing happened. I ran right through the formation and slipped to the outside—I couldn't throttle the airplane back. I decided that I had to go home, but I could not go home without killing the engine. So, I got my gear down—got everything down that I could, and kept increasing the pitch of the prop so it would stall the prop out, and it would not get so much thrust. I finally got the plane slowed down and then had to shut off the fuel and make a dead-stick landing, which I did on a tiny strip in a coconut grove. You cannot misjudge too many times, because if you are falling short, you are not going to make it. If you are too long, you are going to go into the coconut trees.

Anyhow, everything worked out on the landing, and as I was rolling up there on the runway, a guy came running out. He said, "What's the matter?" I said, "I don't know what the matter is with the dang throttle. I just couldn't pull it back." He reached in there and pulled the throttle right back. I sat there dumbfounded. Why couldn't I pull the throttle back? What had happened was that one of the mechanics had left a screwdriver on the plane, and it was jammed into the throttle connection up under the instrument panel.

I had one other big battle out of Munda. We were up one day and had seven airplanes with us—one flight of four and one flight of three. Suddenly, a Japanese fighter sweep came in over Vella Lavella. We vectored around and finally tally-hoed. Somebody called out that there were 50 Zeroes. At the time, because of seniority, Ace Newans was the flight leader.

So we dove into the Zeroes. The Japanese were all strung out in strings of four just like our elements, only they strung out in daisy chains. As you can imagine 50 airplanes milling around in a roughly moving package is like crossing a banded sky. We dove into them four times. As I was "Tail-end Charlie," I was sitting trying to follow the leader and at the same time trying to keep from getting shot from the rear end. After the third time through, I was no longer following anybody. I was all on my own. I came around for the fourth time, but by this time I was sitting up in the sun. I decided to follow the sun line down and join right up on another flight. But what I saw were four Japanese planes sitting there. I trailed in behind one of the wing men—their Tail End Charlie—and fired. As I hit him he started rolling over, and going to pieces. Then I realized that was my mistake because, if I had sat there, I could have sawed on all four of them. But I followed him out. Then, of course, the other three were sitting behind me and working on me.

The Corsair could always out-dive the Zero, so I dove on out of there and came home. When I got home, I found that Curly Koetch, my section leader, had not come back. I felt really bad about that because I had left him in this melee. But he did live. He had been shot down and rescued at sea. He came home about two hours later. Curly was one of my greatest compatriots. He got me out of more problems than I did for him. In fact, he was always saving my bacon.

*Give me a story about that.*

Another time he saved my bacon before we flew to Rabaul. I first had gone for an engine test, because my airplane had supercharger problems. If the rubber seals on the superchargers leaked, you could not get your power at altitude. Anyway, we were scheduled for high cover of a dive-bomber run. The engineering officer looked at it and then asked me if I still wanted to take my plane, and I said yes.

Normally the dive bombers would come in at about 12,000 feet. We would operate up to 18,000 feet on the high fighter cover stacked down to about 15,000 feet. As I headed up on this bomber run, I found my plane was not getting any power. Ironically, it felt like I was flying a bomber, and my fellow fliers like Curly were protecting me, zigzagging over me. We used to fly what we would call a "Thatch Weave," but I didn't even have enough power to weave with them. I was flying straight and level and all these other guys were weaving back and forth. Then the Zeroes started stacking in on us and began to pull up alongside in their daisy chain. Then they would start to fly through our ranks. Unfortunately, we couldn't time their runs. There would be some guys coming through there all the time shooting at somebody.

The last time I saw Curly he was flying right straight up into one of these flights of four Zeroes—pulling right straight up in there. Then he stalled out. His airplane was spinning and coming down—a Japanese plane right up his tail. When I got home, I was really feeling bad about Curly, I said, "I'm damn lucky to be back, but I don't think Curly is coming back."

A half an hour later, this Corsair starts coming around in the traffic pattern and by golly it was Curly. I said, "Let's go out there and see Curly and see the holes in his airplane." Finally, about the only thing I could figure out was that the Japanese couldn't get the proper lead on him. They didn't know which direction that airplane was going. But that Japanese was right up his tall shooting like mad. Curly had spun all the way down from 18,000 feet. The Corsair was not a nice spinning airplane. In fact, spin recovery was not that good. But he spun down to about 3,000 feet, and by the time he got there, no Japanese plane was with him.

Another time Curly and I went barge hunting around Choiseul Island, and once again, Curly saved my bacon. Barges would run at night mostly, carrying supplies for the Japanese. In the daytime they would go back in under the trees and hide along the coastline. The only way you could really get at them was to come in from the sea shooting and pull out at the last minute. On this particular day we found some barges with a 1,500 foot hill behind. We made a couple of runs but did not notice any of them burning. Each time I kept driving in closer and closer. Finally, I got in a little too close and couldn't pull out in time. I ran into the top of a coconut tree. The collision bent the propeller and the aircraft started shuttering and shaking. Here I was sitting a hundred miles from home with a damaged plane. Curly was my section leader, and he comforted me all the way home. When we got home, Curly, the engineering officer of the squadron, changed the propeller on the airplane and scraped all the foliage off the bottom of the wings, and I kept right on flying it.

*Can you describe that friendship with Curly?*

It's hard to. It seems like when a guy gets a bond to you that you can't replace him. You really made the ultimate commitment to them – you trust guys with your life, or your life to the guys. And they do the same to you. It's an important feeling that I don't think can be matched anywhere. I'm sure it's there in almost any combat situation. I know it is among pilots.

Another unforgettable character was Zerky. I had been overseas for about eleven months on my second tour when the war ended. At the time I was Squadron Operations Officer, directing all the flight organization. We figured we would be relieved and would go home. We thought, "We're going to be here for one more month." Just after the war ended, Zerky and several others were flying barrier patrol, which was a patrol between Japan and Okinawa, to ensure hostility actually ceased. They decided that they'd get a little closer to Japan and take a look. When they flew over, one of them had engine trouble, so they decided they couldn't let one person land. Instead, all three of them landed in Japan. At that point we started getting messages back that these guys were in Japan with a bad airplane and that the Japanese were treating these guys royally. (They were the first Americans actually on Japanese soil.) We sent up transports full of mechanics to fix the plane and finally got the men and planes back.

Zerky had been my bad boy, but I loved him because he was a good pilot. So here we sat in Okinawa, knowing full well that we were not going home yet. We had another month to go, or so we thought. One day Zerky came up to me and said, "I would like to go to China." About this time the fleet air wing was running up close to China, but not into China. I said, "Okay, Zerky, I'll keep you off the schedule, but I don't want to know about this." So Zerky and the other guy—J.J. Quinn—went up into China. No sooner had they left, maybe about two days later, orders come in for us to go home, without relief. We began desperately packing and trying to get everything squared away. Meanwhile, I was sending messages by way of the Navy up into China to try to get Zerky to come home, but I could not reach him. I think he went to Shanghai. Finally, I confessed to the skipper. I told him, "Zerk's up in China, and I can't get him back." It was clear Zerk wasn't coming, so they let us go home. A few days after we left, Zerk came back to a deserted base. He was held in Okinawa for many months, but the military cancelled the court martial because of lack of witnesses and stipulated that he resign from the Corps. Zerk was a character. The last time I saw him he was still pumping gas down in El Segundo.

*Give me some other ironies of war*

I was overseas 27 months in WWII in two tours. Neither Curly nor I had family to tie us down, so we decided to go right back overseas. In two months preparation we became part of an outfit that trained down in Pendleton, California. During this time we had all these junior bird men there. I was a combat pilot from my first tour, and I thought I was pretty hot. It was here that I met Zerky, who could fly rings around me, so I kept him in my division when we deployed. We never saw an enemy airplane in the air this time.

*You had how many kills?*

I had three and one that was a probable. Anyway, I felt that I was good. I was more accomplished at tearing up our own aircraft. When we were training at Camp Pendleton to go overseas again, we had a hospital Cub airplane, which had a swing-back in it where you could put someone. One Saturday when everyone else had gone to town, I made a commitment to the squadron barber that I'd take him for a ride. I seated the barber behind me on an ammunition box. We had decided we were just going to go beat up the airfield. I had never flown a Cub but I thought I could fly it, so we took it up. We were really beating up the airfield, doing chandelles and barrel rolls and loops and coming down and strafing the field—all sorts of things. At the end of that, the idea was that I was going to come in and land and pull right up to the end of the chocks where I had gotten the airplane, which had been parked alongside some Dallas huts. I was going to come steaming up, hit the brakes, spin her around sideways and sit the plane right there in the chocks.

The plan was working fine. I came steaming in and made a real air show of it, including a nice landing. Then I came taxing up like mad towards the Dallas Hut. I tried to stop and pushed with my heels, instead of my toes, because this plane had heel brakes. I pushed and pushed on my heels but nothing was happening. I pushed a little harder. Finally, I pushed so hard I pushed the canvas back right out of that seat and slid backwards and fell in that barber's lap. The damn airplane didn't stop! I had gotten turned a little bit when I hit the side of the building. The propeller broke and was flapping up against the side of the building. I looked down to see what had happened. It just happened that this airplane had a fire extinguisher with the spout going one way and the handle the other. My heels were actually bearing down, not on the brake, but on the fire extinguisher. When Monday morning came, the squadron flight surgeon came in and was really mad. This was his airplane that he gotten flight time in. He said "Wes, you been flying too much?" I said, "No, no Doc."

I think the only thing that saved me in terms of my career was that they needed someone to go back overseas. Actually, though that's the only accident I ever had that I was responsible for. I had a number of accidents, but they were not caused by pilot error.

One day we were sent out on what they called "Roger, Peter Patrol." This was a circling patrol around a group of ships to intercept kamikazes as they came through. The kamikazes were really blowing things up about his time. It was 1945. You would have a pre-briefed direction that you approach the fleet in. I came in one day and made my proper entrance corridor. I called in and everybody agreed that I could come on in. As soon as I headed in, though, the whole damn fleet started opening up its guns on me. I promptly reversed direction and went out in the ocean a ways and called in again. Then I flew back in again and the same thing happened. Twice everybody in the damn fleet was opening up on me. They were so trigger-happy. All it took was for one guy to start to shoot and everybody would shoot. It got to be pretty exciting.

*So you got out of there then?*

No, they finally let me in. But I never intercepted any kamikazes. None ever came while I was on station. We had some other strange experiences, though. When we were training between tours, my squadron leader was named Frazier. They called him "Fearless" Frazier. He was a stuttering person. But any time that he really wanted to say something he didn't stutter one bit. He come out of the South Pacific also and had a very enviable record in Squadron 122, I think it was. But he had an accident there at Pendleton. He came in the traffic pattern. We used to have inflatable boats in our seat packs, and these things periodically would inflate all by themselves, something like an airbag nowadays. To take care of this problem, we had a little strap of sharp steel that we actually taped to the cowl of the airplane, so if you had trouble you could reach up there and grab that sharpened steel and jam it into this boat. It just happened to Frazier at the wrong time right in the traffic pattern. That was one of the hazards of flying. The boat inflated under him at 800 feet, and it pushed the stick forward and he went right into the ground and was killed.

*It's always tough seeing your buddies go down, isn't it? Did you go to the funerals?*

Yes, we would go to funerals. I buried a few. The most touching thing I ever did was to bury some of these guys. At Vella Lavella we had a lot of operational accidents, in addition to the losses from combat flying.

We were losing a lot of pilots. The weather was bad and we had take-off accidents, even. Tires bust. Strange things would happen to airplanes. Jesse Leech and I were the new boys in the squadron when we had a kid crack up on take-off and burn. The men got the corpse out. Then Leech and I were selected for the burial detail. It helped to put these people into the ground real quick because of the tropical climate. So you would do a hasty burial.

We were on the truck and had the body in a sheet. Jesse and I were on the sheet, lowering this sheet and the body into the ground. Suddenly, we bent the sheet and compressed the body so that the air exhaled from him. That was a pretty wrenching thing for me. A burned corpse and it is groaning while you're putting it into the ground.

*He sounded alive?*

Yes.

*It's like he was talking to you.*

Yes. Later on the burial details would come around and dig up the remains and send them back and everything. During the height of the operation you don't have time for that.

*You see some pain. You see some incredible things, don't you?*

War is a brutalizing experience, even for aviators.

*Did you have some carryover after the war, some nightmares?*

For a long time, I didn't. I used to think I was kind of hardened to that stuff and used to think that I had an inner faith in myself—the feeling that what will be will be. Some of these guys would drink at night and start crying in their beer and talking about those dear, departed souls. I never did that until just a few years ago. I started seeing ghosts. I don't mean literally seeing them. Thinking about them. They were there.

*Tell me a little more about that.*

I can't really tell you. I just think you begin to realize the futility of a lot of things. You do get to bring forth these memories. I've had two mid-air collisions in my life. In all cases, everybody else was dead. You wonder about those things. I had two bailouts in the Philippines. The first bailout in the Philippines was when we were sitting in Green Island. The second time we went overseas we went back into the Solomon Islands and the Green Islands and were immediately called up to the Philippines because the naval commander wanted Marine air cover rather than Air Force. We were supposed to take a squadron into Peleliu. As I took off from Peleliu, I had hydraulic trouble. Rather than go back to the island, I wanted to go ahead and make sure that I got into the Philippines with the rest of the guys. I figured when I got to my destination I would blow the landing gear down, even though I had no hydraulics.

The way we would make these long over water hauls was to fly with a bomber, which would act as the pathfinder. We would fly four Corsairs on each side on the wings. On this occasion, we had to fly into a typhoon. The weather was really, really wild. You'd see an airplane next to you suddenly go up above you, then down below you. It was really rough flying. In this, one of the guys on the other side had a collision and one airplane went down. The other guy opened his canopy and was getting ready to go jump. But he figured it needed to continue to fly, because the plane wasn't that badly damaged. He and I were hanging back, because we couldn't keep up with the flight.

When we hit landfall someplace in the Philippines, we didn't know where we were. It just happened that the Japanese were giving us a bum steer. They were on our frequencies and rather than directing us into a destination, they were giving us the wrong way to go. We would fly into landfall, but the bombers were telling us fighters to go ahead and ditch, that we knew where we are.

They didn't know where we were. We went in and the first thing that happened was when the landfall came into sight, I decided that this was my chance to get my gear down, so I started pumping the gear down by hand with the remaining residual fluid that was there. Lo and behold, there was no landing field there. Apparently, the bombers didn't know where we were, and then the Japanese started steering us right back toward San Francisco.

We flew back out to sea for about a half-hour. Normally, a run from Pelelieu into Leyte would take about three and one-half hours, but we flew around there for about six hours and hadn't gotten anywhere. We finally come in and hit land again. This time the guy that lost the canopy decided to follow the coastline, and he was the only one that got to the right field in Samo. The rest of us went over the mountain ended up at Leyte where the amphibious landing had just taken place. As we came in, we saw the military was still ferrying stuff in from the beach in amphibious tractors and landing craft and the ground troops had secured the field. It was the normal amphibian landing scene. A big mess.

A couple of fighters landed on the field and lined up down there. By the time I came in to Leyte, I had been out about seven hours, and was running out of fuel. In fact, I was running on fumes for the last five or ten minutes. I tried to blow the gears down. One came down but one didn't. Finally, with one gear down and one gear up and running out of gas, I wanted to land, but the people on the ground were not going to let me. They said, "You're not going to crash land here. Go down to the beach." They didn't want me to run into those other airplanes that they had down there on the ground. They insisted, saying, "You have got to go crash land on the beach." So here I was in the air trying to figure out what to do. I guess I must have sounded pretty excited, so some guy came up on the radio and said, "You better bail out, Sonny, airplanes are cheap." So that was my first bailout.

This was going to be a deliberate bailout. So I thought, "Well, I can't swim very well, and I don't want this airplane running into a lot of people out there in the bay." I decided to trim all the left tab into right tab, and when the airplane started rolling over, I decided I would stand up and go over the side. Everything worked out just as I'd planned. I came across the beach at about 2,200 feet, headed out to sea and cranked the tab in there, stood up and when the airplane was upside down, I slid off the tail end of it. From only 2,200 feet you do not have to wait long to pull the cord, so as soon as I was clear of the airplane, I pulled the cord. I had a brand-new, personal .45 pistol that a friend had given me, not a service model, which was in my shoulder holster. It fell out and hit me in the mouth and fell into the ocean. I was treading water there when I pulled on the knobs of my Mae West but pulled too hard and they came off. I figured I would save this chute, so I had the chute in one hand and was swimming along when an amphibious tractor came by and tossed me a line. Now I had the line in one hand and the parachute in the other hand, and the parachute was like a big sea anchor. The tractor kept pulling me and because I have this sea anchor in one hand, I had to let loose of the rope to the tractor.

When we separated, the tractor, a big, cumbersome beast, went out and slowly turned around. Finally, the driver made another pass by me. A sailor jumped in the water and said, "You take the rope and I'll take the parachute." Then they hauled me into the beach. Of course, all my clothes and personal belongings were in the plane, except for the things I had on my back. As a consequence, I had to live like a Seabee for about a month before I could get another flight suit again. We didn't have enough supplies or anything. I flew in a Seabee working uniform for about a month.

A month later, I had gotten over to Samo and we were to make predawn takeoff to cover carriers. After the take-off, I made a 180-degree turn to enable the wing man to join up. He seemed in position, just a little wide. I went on dim lights as we had a pre-briefing. As he closed in, his prop chewed into the right wing and tore out the bottom of the airplane. He went in and blew up. My plane just hung in the sky. With open throttle and control stick broken off, I had time to think. You had to pull pins on two sides and push the canopy back to throw it off. This time instead of jettisoning the canopy, I used my left hand to push the canopy back, which ultimately saved my life. Since I had been out a month earlier I thought, This airplane's going to snap roll to the right. When I start getting away from it, I did not want to be on the inside of the snap roll at 800 feet of altitude, so I decided to get out on the left. I pushed the canopy back, went out the left side and the damn nose came up and the airplane snap rolled to the left. Now I was sitting there on the inside of a snap roll at about 800 feet. The thing that saved me is I could claw my way out. I could reach around and grab that canopy and pull my way out of the inside of the snap roll. Immediately, I pulled the chute and the last thing I remember before I hit the water was seeing the chute blossom up between my legs in the night sky. It popped and I was in the water. It happened that I was about a half-mile or a mile off shore, but fortunately the water was very shallow there and I could stand up. Unfortunately, I couldn't walk because my leg was hurt, so I took out my pistol, which had tracers in it. I heard natives or somebody around there in boats so I thought I would give them the international distress signal. I fired three tracers. Then everything went quiet. Nobody was moving. About a half an hour later came dawn. I saw a guy rowing his wife down the beach. He was taking her to the doctor. He picked me up and took me in to shore. My leg was hurting pretty bad. We had a baby doctor who was a squadron physician. He took one look at my ankle and said it didn't seem to be broken but that it had a real bad compression bruise. He said, "Here, I'll give you some Novocain." He took a syringe and a thing like they used to vaccinate cows with and jammed it down between the bones in my ankles and shot me full of Novocain. Then he said, "Wes, get up and walk." So I got up and walked. That was the last I walked for about two weeks. When that drug wore off, I was on crutches. I couldn't stand up for weeks.

The baby doctor was a strange guy. One of the common complaints of fighter pilots is hemorrhoids, because we pull a lot of G's and load those blood vessels up in our lower extremities. If they inflate and over-extend, you get a knot on them. So I developed an external hemorrhoid about that time in life. It was bothering me and the doctor was going to operate on it. Here he's used to looking at these things you know. He had me get up there on the couch, and he shot the infected part with Novocain.

Next, he gets in there and cuts open that damn vein and pulls out the clot, which I could feel him slipping out. Then he sewed it back up. In about three days I was back in the cockpit. He was a very experienced guy.

*That's all it took to get you back.*

Yes, I haven't been the same since.

*What about these ghosts? When did these come?*

I saw "The Fighter Pilot's Story." To me, that television program really hit home. There are so many terrible experiences. I still look to see combat footage of the Corsairs flying. I think even before that program I began to feel some of the emotions come back. Maybe it was when my grandkids started getting to be young men. Now, as for my own kids, I never advised them to get in any of the services, and they didn't. Sometimes neighborhood kids and everything would come around to hear sea stories and want to know if they should go into the service. I would always ask their mothers if they wanted me to steer them that way.

*What ghosts would you see?*

I see the ones who are gone. I see Willy; I see Leech, Huckleberry, Joey Craig and Gay. Gay was the young guy that collided with me that I just told you about on the predawn take-off. Here I was the old guy by this time with this young pilot flying my wing. I'm not sure that I really didn't cause the accident. I think what happened was that I went on dim lights when he was not looking. At the time he was flying on the brighter navigation lights and then all of a sudden when they became dim, he must have thought I was further away and turned into me. I often wonder if I caused the accident. I suspect that someday I'll find out.

*You expect some day you'll find out? You mean that you'll meet up with this guy somewhere?*

There's an old line that says, "There are no fighter pilots now in hell." And the other guy I had the collision with that I see the ghost of is Charlie Kerr. This was a stateside accident. At the time we were out on a pre-briefed training. Four of us would be targets and four of us would be attacking. We were doing overhead runs on this Division. As I went through attacking the lead plane, one of their planes swung in and we collided. It just so happened I had attended Charlie's wedding just about a month earlier. Sadly, he and I had gotten in the same airspace. Of course, I had bailed out twice before, so I knew you had to get away from these airplanes when they were going down. I got out immediately and made it out. Charlie's airplane was gliding and seemed to be under control, so he elected not to get out. Then at the last minute, it stalled out and cracked up. It killed him. Here I am, the guy involved in the accident going over to tell his wife he's dead and to apologize. It's a tough go.

*Describe that.*

You can't describe it. I went to Cherry Point, North Carolina, to tell her how the accident occurred and how sorry I was. But it was gut wrenching because we were close friends, and I had just been the best man at the wedding.

*How long did you spend with her?*

About a half an hour. I have never seen her since.'

*What did you talk about?*

Just how it happened, how sorry I felt about it.

*What did she say to you? Was she angry?*

She wasn't angry. I think Marine wives know in their hearts what the game is.

*What will you say to these guys when you see them?*

The ones that I think I might have been responsible for—or could have been responsible for—I would say how sorry I am. And the rest of them, just "hello"—tell them sea stories.

# A Co-Pilot's Tale
## Clair Tracy

Clair Tracy flew thirty-five missions to places like Cologne, Berlin, and Muenster as copilot in the 351st Bomber Group, based at Polebrook, England, near Peterborough. Tracy grew up in Aledo, Illinois. On November 10, 1941, he went to Camp Grant in Rockford, Illinois. There he had a choice of the Army for one year or the Air Corps for Corps Technical School at Sheppard Field, Texas and Wichita Falls. After completing school, officers encouraged him to be transferred to a photographic outfit at Felts Field in Spokane, Washington, where he worked as a P-38 mechanic for several months. When the requirement of two years of college was dropped, Captain White recommended him for cadet training. While waiting for class, he went to an old Conservation Corps camp east of San Diego where he and his colleagues would run around in the woods and take three or four-day passes to Los Angeles. Then he went to Santa Ana for basic and Blythe, California, for his primary training in Ryan PT-17s.

But for luck and fate, the Tracy family could have been the material for their own *Saving Private Ryan,* in the sense that four brothers all enrolled in the military for WWII. Vernon was a bookkeeper at Davenport, Iowa and became a navigator in B-24s; Glenn earned his pilot wings and received his commission toward the end of WWII; and Virgil enlisted in the Navy in 1942 and returned several times to the states when aircraft carriers he was on were damaged. Then there was Clair, the copilot, who survived as did his brothers and offers his perspective from the second seat of a B-17.

We sit down at his spacious above-garage apartment, overlooking a rolling, snow-filled field of canola stubble. Clair looks military in his brown pants and slightly lighter brown shirt. He seems anxious to talk.

*How did you get to Europe?*

We left Lincoln, Nebraska, in a thunderstorm at about midnight and flew our own B-17. We had put our crew together in Alexandra, Louisiana. Moline, Illinois, was on our route, and that was only thirty miles from my hometown. At this time, I was flying co-pilot, so I said to the pilot, "Can I have the plane for a while?" He looked at me with a funny look. I then said, "My hometown's over there." He said, "You've got the plane." We were flying at about five thousand feet, so I just put the B-17 in a glide so I could get down nearby my hometown. I figured I could recognize it because it had these globe lights on Main Street. I glided down over town and opened up my throttles at four in the morning. I had previously called Lila, my wife, and told her not to sleep too soundly that night. She had gotten the hint. (We really couldn't say over the phone what we were doing.) Anyway, Lila and her sister were out in the yard. She said the whole sky lit up when the exhaust fumes of gas caught fire after gliding down there. After that, the ball turret gunner, Jack Mack, said, "Why not fly over my hometown?" I said, "We can't show preference here." The pilot gave the okay, so we flew over Jack's hometown of Valley City, Ohio. From there we went to Manchester, New Hampshire, and then to Gander Field, Newfoundland, then over Greenland's ice fields at midnight and on to Iceland and then to England.

*Give me a memorable mission.*

We flew missions to Berlin, Cologne, and Muenster—among others, but it was late in the war. The toughest mission was to Politz, near Russia, where we bombed an oil refinery. This was the longest mission. It took ten hours and was the worst. We were hit by flak and lost an engine over the target, so we had to drop out of formation. In addition, one of our waist gunners got flak in his knee. It was a close call. Conditions were so bad that when we first flew in to Politz, we could not see the sun—there was so much flak smoke. The air was alive with it. That was the first time I felt the ship bounce from flak bursts. After we got back home to England, the debriefing people were nice enough to tell us that at any one time over Politz there had been four hundred flak guns trained on us.

Next, we headed back close to Sweden, across the North Sea. We could have gone to Sweden for the balance of the war, but chose to try to get back to England for more missions. Several times, we wondered if we made the correct choice. We were flying over water to avoid flak. At one point, we saw we were getting low on gas, so we started dumping anything heavy. The Germans had shot out part of our radio. The pilots usually had some radio contact, but the only radio working was back with the navigator and radio operator. I would say that for the last twenty minutes all the gas gauges showed empty.

I was riding co-pilot and the pilot started circling, trying to get permission to land in the first available field in England. Pointing to the gas gauge, I said, "Wheels down. Land the thing." As soon as we started downwind, we had an engine quit on us. Now we only had two engines. On final approach, another engine cut out. When we landed, we had only one outboard engine. As a result, we couldn't even taxi off the runway, because when we gave it the gun the plane only wanted to go in circles.

The next morning the pilot said we wanted to take off for our home base. I told him, "On this short fighter runway I'm not sure that we can make it." He said, "We'll make it all right." The trouble was we had a crosswind into that one bad engine. When you have that situation and you get the speed up and down the runway, the plane wants to twist on you. I didn't want us to attempt to leave then, but the pilot wanted to and he was in command. We just barely got over the hedge. I think of all the missions I flew that was possibly the most frightened I got was on that takeoff.

On one mission, we flew "Tail-end Charlie." As you know, one plane always flies behind the others in case somebody drops out. This occurred on a mission where I was checking out a first pilot. Anyway, we saw a plane's engine smoking up ahead. The new first pilot was wondering what to do. I told him the procedure was for the guy behind him to fill in when he drops out. Suddenly, the smoking plane dropped out of formation, but the guy behind him didn't fill in his spot. I said, "Let's go over there and fill in where the smoking plane was. First, make sure that we're clear of any other planes." We slid into the vacated spot, which put us in the center of formation.

*Did you get in trouble for that?*

I thought we might, but our superiors never did call us on it. One pilot I flew with liked to drink a lot. This time I was flying with another crew, not my own. It seemed the pilot had had a bad night the night before. After sitting there for some distance and getting up to elevation, he turned to me and said, "I'm tired. I'm going to take a nap. You've got to fly the plane." I did and that was the only time I took over without the first pilot being in charge. After a time, he got to feeling better, and by the time we reached the target, he was able to take over and fly the plane to the target.

*Was that fun being in control?*

Yes, and that gave me confidence. Later on, I was able to at least check out as a first pilot.

*Explain the distinction between the pilot and copilot.*

The copilot was there to assist the pilot, but I probably flew the plane half the time, once we got up into formation.

*Was there any kind of attitude toward the copilot that you experienced?*

I didn't realize it until we were about three-quarters through our missions, but I found out that other first pilots were dividing their first pilot time with their copilots—splitting it fifty-fifty. I went to our first pilot and said, "Here's what these other guys are doing. Now that I'm about out of the service in England, I need to start thinking about what I'm going to do when we complete our missions." I added, "I wouldn't care, except that it would help in the future to have some first pilot time." He said, "Oh, I didn't realize they were doing that. I'll see if I can get some of this time brought over to you." Later he came back and said he couldn't, but arranged for me to get some split-time on future missions.

As it turned out, it didn't mean anything to me. When I got back to the U.S. after the war, I was one of the first ones terminated from the service. I figured I would end up going to Japan, but instead I was released. The reason might have stemmed from the fact that there were captains who had two tours of duty who had to go to Japan on B-29's. I felt fortunate, really. Probably it helped me not having that much first-pilot time. Someone said, "We don't need you as an instructor, because you don't have that much first-pilot time."

*Did you get spooky at about thirty missions? You had five to go.*

We were lucky. I only saw one German fighter and a P-51 was putting him on the deck. We didn't get there until September of 1944. By that time, the German fighters didn't have fuel. The thing to remember though is that you didn't have to go to Germany to have trouble. There was always weather coming back into England. It was hard to find the runways. I remember we came back at night once and were trying to find the runway. The pilot said, "I don't see it." I said, "You think we ought to go around?" He said, "I think I see it now." I said, "I don't." A few seconds later, I saw the runway light and I said, "I see it. Land it." I called out, "Air speed," and lowered the wheels and flaps. The pilot made a great landing. All the crew cheered, and we were all relieved to be on the ground, even though it was another strange English airfield.

*Any wild and crazy things happen over there?*

We had one or two days before this particular mission. We went out, flew the mission and were starting back when one of the guys said on the intercom, "I've got to go to the bathroom." But the B-17 had no provision for diarrhea. We had these empty chaff boxes, which contained shredded aluminum to mess up enemy radar. They were big boxes—about three by six. I told the guy, "There's a chaff box back there. Use the chaff box." Then two or three of the others had the same problem. And pretty soon, I had to go, too. I went back, used the chaff box, and then climbed back in the cockpit. Pretty soon, the waist gunner said, "I can't stand the smell. It's terrible." I said, "Open up the bomb bay and throw it out." He did just that. All of a sudden, the ball turret gunner said, "What's going on up there? I can't see a thing." I said, "You might as well come out of there." We were out of danger by then anyhow. I heard later after I got back that there were several crews that had that kind of experience. Later we figured out we had had some bad water from flooded water reservoirs.

*Did you have any forced landings?*

Right after I soloed, my engine to my Ryan PT-17 quit due to a broken throttle control near the engine. I was over the Colorado River Valley. I thought I had myself set up for a perfect landing without power. I had to take whatever glide pattern I had at the time. When I landed, I barely touched the first bank of an irrigation ditch, and the nose of the plane went over and hit the second bank. The plane then flipped over on its back. I wasn't hurt, except that I pulled a few muscles, but luckily, that was all. At that time, you were supposed to stay with the plane and wait for someone flying over to recognize you and waggle their wings. No one came by so after about a half hour I went down to a farmhouse to call the base to let them know where I was, but there was no phone. Apparently, someone had seen my crashed plane and reported it, because when I got back to the plane, the base commander was there. He was a major, and he said, "Are you hurt soldier?" I said, "No sir." He said, "Well come to attention when you're speaking to an officer." I was surprised, but I figured it was shock treatment designed for me to get upset with him and not the accident. Back at the base, the captain told me to go back to whomever had soloed me. The instructor took me up and said, "You land this thing unless I take over." So he gave me about five forced landings. I went out over more than one barbed wire fence, and just as I was about to set the plane down, he took over. After about five of those, the instructor said, "Let's go back to the base. Son, you just had bad luck. There's nothing wrong with your procedures."

*Did you lose any friends in the war?*

I lost one friend. I heard through my cousin in Springfield, Missouri, that my friend was shot down and killed in the war. He was young, probably nineteen at the time. I was twenty-three years old, but most of our crew was eighteen or nineteen years old. Anyway, I had flown with him in basic training. We flew "The Vultee Vibrator" or the B-13, as it was called. It was a two-place plane on final approach. The guy who wasn't flying would sing "The Old Rugged Cross," and we would try to touch the wheels down on the last word of the song. It took some of the tension out of landing, and trained us to be more accurate.

*Was there a split between these younger guys and you?*

They would usually come to me to borrow money. I was trying to save money to get my life going when I got out of the service, and of course, they were not earning what I was as a copilot. At the time, I had authorized the government to send so much back home each month, but I still ended up with seventy dollars of discretionary income. I was making three hundred and seventy dollars a month with flight time. Even with seventy dollars, I had extra money available, so I could loan them money when they asked. I never lost a dollar from any of those guys.

*They all paid you back?*

They all paid me back. Usually, five dollars would go a long way in England then. On a night pass, I might have had thirty dollars out at one time. The officers used to have to censor the enlisted men's mail when they wrote letters home. Our crew was going to the pilot for that. Pretty soon, they started to come to me. I would say, "Isn't Sully doing a good job for you?" They said, "He reads our letters. I don't want him reading what I am writing home to my wife." So I said, "If I can trust you guys back there with those guns to protect me while I'm flying, I can sure depend on you when you write letters to your wives. If you want me to do it, I'll censor them, but I'm not going to read them." After that, they came to me, and I signed off on them. Pretty soon, they weren't coming to me. I wondered what was going on. "Aren't you guys writing your wives anymore?" I asked. "Yes, but we learned how to write your name." I said, "I'm not going to take the blame if you get me in trouble."

*Any other pilot stories?*

We had one copilot over there from South Carolina, who liked to gamble. He had served two tours over there and wanted to put in for another, because he was making money gambling. I heard he won thousands. But he was scared to death of flying. I don't know how the first pilot put up with it. Someone said, "He'd get in the plane and as soon as it got into dangerous territory, he would grab a flak suit and pull it over his head and just sit there and ride."

*He wasn't willing to gamble his life, but he liked to gamble his money.*

Yes. (Laughs)

*I interviewed a pilot who did not seem to think religion was that important over there.*

I always carried a Bible with me. I felt there was nothing I could do except do my job and the rest was in the hands of the Lord. That's the way I felt. Anytime we went in on a bomb run, from the initial point (IP), I always said a prayer. I said, "Lord, it's in your hands." I would pray that I was doing His will, not ours. That was pretty much the way I approached it. That's also probably why I didn't have any real fear—or didn't get to the point where I would lose it emotionally.

Some people did lose it. This involves a story. I used to fly with Sidney Townsend. In fact, Sid and I crashed a plane in advanced at Douglas, Arizona, during a night takeoff engine failure. It so happened that Sid and I landed this old bamboo bomber, a Cessna At-17, and when we did, both engines tore off the plane. Just after we landed, we looked out and saw them sitting fifty feet from the plane. Anyway, not too many years ago, my wife and I were traveling in Louisiana, and I called Sid up in Alexandria, Louisiana, where we had formed our crew. The phone rang and someone answered. I said, "Is this Sid Townsend." He said yes, and then I said, "Well, there are a couple of engines lying out in the desert in Arizona, and they say you're partly responsible for putting them there." He said, "Who is this?" and I told him.

My wife and I went over and visited him and during the visit, he told me about his first pilot, a fellow named Tulley I had known earlier in basic training. It turned out his plane got shot up really bad and Tulley froze on the controls. Sid had to get the navigator up to help get the guy out of the pilot's seat. It took two of them. Then Sid managed to take over the crippled plane. Sid took over the crew from then on for the rest of their tour of missions.

*The fellow I just mentioned suggested that few pilots prayed before they went out on a flight.*

I would say there was a lot more religion than what people knew about—or people will admit. There was a lot of prayer. Plenty of prayer also went on back home, and I believe that reached us over there.

As of October 1998, five of the original crew were still alive and able to meet at a reunion in Kansas City, Missouri. They included Jack Mack (ball turret gunner) Valley City, Ohio; Dale Rodness (waist gunner), Minneapolis, Minnesota; Joe Heffner (bombardier), Colorado Springs, Colorado; Herb Sass (engineer), Belton, Maine; and me, Clair Tracy (copilot) Spokane, Washing.

# Vietnam Rescue

## Dennis Hague

Dennis Hague played a role in one of the amazing rescues of the Vietnam War, flying cover for Bernard Fisher who won the Medal of Honor for his efforts.

Hague is a slim man in his fifties. When he gets to talking about flying, his demeanor changes, and he becomes animated. Clearly, he has lots to talk about. Count on him to be thoughtful and reflective.

Dennis Hague went through ROTC at the University of Idaho and entered pilot training as a second lieutenant. He was chosen for interceptor school—F102s and F101s. His first operational assignment was in Great Falls, Montana, in an air defense squadron, flying 101s. His assignment to Vietnam was served in the First Air Commando Squadron. After Vietnam, he returned to Fort Walton Beach, Florida, where he instructed A-1s for a year at Hulbert Field—the same base on which they trained Doolittle's Raiders. He points out, "The actual runway that they used to train on, we used as a parking ramp when we had live ordnance on our airplanes. We'd just say, 'They're parked over on the Doolittle.' My God, you talk about bravery and dedication, think of those Doolittle guys." After that experience, he joined the Washington Air National Guard. His affability and intelligence, verve and wit—and dedication—have served him well. He is now a general, but there is nothing stodgy or haughty about him. We gather at the round oak table and begin to talk.

*You flew the Sandy in Vietnam?*

Yes, "Sandy" was kind of a generic name given to the A-1 Skyraider. "Sandy" was the call sign for the birds that primarily supported air rescue. When pilots were shot down, the Skyraider, because of its performance and durability and its firepower, would normally escort the helicopters—the Jolly Green Giants—in to pick up those pilots that were down. The mission was to keep the bad guys away until the helicopters could get in there. The call sign for the unit I was assigned to, even though at times we had a "Sandy" mission, was "Hobo." We were the First Air Commando Squadron.

*The Sandy was like the B-57 in the sense that few people flew them. Isn't that true?*

From my vantage, for the people that were my age to return to a propeller-driven aircraft, was a really unique situation. A lot of the folks felt that we were into the jet age. In fact, I had come out of supersonic interceptors when I got my assignment to go to the Skyraider. So, to go back and fly a prop-driven aircraft, where you could open the cockpit and let your scarf blow in the breeze, that was really unique.

There were two squadrons of Skyraiders in Southeast Asia. I was in the First Air Commando Squadron—those were the Hobos, and there was the 602nd Air Commando Squadron—their call sign was "Spad." But the "Sandy" call was a call sign assigned the air rescue.

*What's the weirdest response you got when somebody saw you were flying a prop-driven aircraft?*

I think some people thought it was a step back. I remember one thing that came up. As you upgraded in aircraft in the Air Force, you pick up a new commitment. And the Air Force was saying, "If we're going to train you in newer aircraft, we want you to stay with us longer." That was one of the things that was somewhat ironic—that even though I'd gone from an aircraft that was capable of almost Mach 2 to one whose top speed was 400 knots—you had to be screaming and going straight down to get that speed—the plane was considered an upgrade.

*What about its performance?*

Well, the thing that was remarkable about it was its simplicity and its durability. From a pilot's point of view, you felt very secure in the thing. You had a large engine in the front of you to protect you. Also, the reciprocating engine could take damage that a jet engine could not. A jet engine, when it would take a hit, had the capability of self-destructing. You knock off one turbine blade, then it knocks off two more. Those knock off four more. Pretty soon it's just a spool spinning back there. The A-1 was effectively an engine and guns with a lot of steel around you.

It will get you home. You could stay on station a long time. It could carry tremendous firepower. We could carry the same tactical load with a single-engine Skyraider that was carried by the B-17s in World War II. And our ability to stay on target—stay with the troops for a long time—was something. Then the fact that it was slower lower. The ability to precisely deliver the ordnance. I was on an air strike where the U.S. troops and the North Vietnamese were in such close contact that the Forward Air Controller, who was going to put in the strike, said the only air he would accept were American-flown A-1s. It was that close. The airplane and the pilots were held in pretty high esteem. Part of it was just simply the performance of the aircraft. You know, if you go screaming through there at 600 knots to drop a bomb and you even blink your eye, it throws the thing one-half mile off. Whereas, napalm and a lot of the things you'd use for close air support meant a low-angle deliver at 225 knots—about top speed.

*What was the toughest transition from the supersonic interceptors to the Sandy?*

The thing is that the jet engine—or jet aircraft—effectively has no torque. But, if a propeller is turning one way, the airplane wants to turn the other. Now this is offset by adding a rudder. Well, most jet pilots thought the rudders were a comfortable place to rest your feet while you fly—you didn't have to use them. But you could always tell a Skyraider pilot because one of his legs—the right one—was overdeveloped. It got that way just making sure you held the rudder in. The engine was about 2,700 horsepower, and had a four-bladed prop that was over eleven feet from tip to tip. If you firewalled it, and brought in the power, and did not add the rudder, the airplane had the capability of flipping right on its back—called a "torque roll." So that was the biggest thing. Then, of course, in a supersonic fighter you could simply reach over, stroke the afterburners, and go to 35,000 feet just as quickly as you wanted to. But the Skyraider did not have that type of performance.

*One can imagine that somewhere there must have been a macho guy, like Tom Cruise, who really couldn't stand to step down to a plane like the Sandy.*

I think that Tom Cruise or any of those fellas would have been involved in the mission. There was not a Skyraider pilot who felt that way, I'm sure. The best people to ask would not be the pilots that were involved in the war, but to go out there and find some grunt, some foot soldier, who at some time was really locked up tight and needed help. That might be my ego playing, but I think the Skyraider pilots were held in very high esteem. By everybody. I think a lot of guys would have given their hats to fly Skyraiders—I really do.

*How do you remember your first rescue?*

The first successful rescue? I was involved in several rescue attempts. In one, the pilot was already dead—the bad guys had gotten to him before the helicopter came in. The thing I remember most was the bravery of the helicopter pilot. We were scrambled to go over there near the Cambodian border—there was a pilot down. He was in an observer aircraft that had been supporting a ground flight between the bad guys and U.S. troops, and he got shot down. We were scrambled off alert to CAP the rescue – Combat Air Patrol. And we went over and, because of the speed of our aircraft, got there before the helicopters. You could see where the aircraft had crashed. It was like maybe ten football fields put end to end, and the bad guys were on one sideline, and the good guys on another, and you could see the Forward Air Controller's crashed plane. You could see his airplane. We hoped he had made it to one of the tree lines. We got over there, got on an emergency frequency, and tried to get him to answer us but he didn't. Finally, the helicopter showed up, and the pilot said, "Has anybody seen the pilot?" And we said, "No." And he said, "Well, he won't respond, and he won't show us smoke." (One way you could tell somebody on the ground, where you were is for them to throw a smoke grenade, and then they could find you.) But, the problem with that was, everybody could see the smoke—good guys, bad guys. The helicopter pilot said, "Well, he's not going to show himself until he sees me." So, he says, "I'll land." "Anybody see him?" "No." So he picks up, goes a little further down, lands again. I guess what I'm saying was that there was this helicopter crew that was willing to risk their lives, and I mean, it was wild city down there. I know that was their job. But, anyway, they finally did, later on, get over there, and the bad guys had already gotten to him and killed the pilot, so we didn't get him out. That was the very first one.

*Then you flew with then-Major Fisher the time he won the Medal of Honor.*

Yes. The Ho Chi Minh Trail—or portions of it— (actually it was a network of trails) would come down just north of the DMZ—between North and South Vietnam. It would swing over into Laos and come down the valleys. The valleys in that part of the country ran almost truly north and south. It would come over into the A Shau Valley and, using that, the North Vietnamese would infiltrate down—or slip back in. Though A Shau Valley was actually in South Vietnam, the North Vietnamese would go over to Laos and come back into Vietnam. There were several routes like that. Normally, the U.S. would put Special Forces camps in there to try to block these operations—there was a camp at A Shau.

The weather had been very bad, and later we found out that some folks thought this would be the first introduction of North Vietnamese troops into South Vietnam. Most of the time before that, there would be a few of them, but it was mostly Vietcong that were involved. Anyway, the North Vietnamese came down and surrounded that camp, dug themselves trenches, then attacked the camp.

It was on March 9 that Major Fisher led a supply aircraft down into the valley. The camp was running out of ammo and supplies. Somehow, he was able to find a hole and lead this C-123 down into camp and drop them supplies. That was his first heroic act. I think for that he got a Silver Star. It was very, very difficult. Dark, bad weather, clouds, ground fire, down below the mountaintops. A very brave guy. But the rescue itself took place the next day, March 10, 1966. This camp had been under attack for about three or four days. The night after the first mission, Major Fisher told us he thought we were going to lose that camp—the weather was too bad, and we couldn't get any air in to support them. There were about 2,000 North Vietnamese regulars who had surrounded the camp, greatly outnumbering it. That night we thought the camp was gone.

*You thought it was gone.*

He did. I hadn't been near it at the time. We heard about his mission and, of course, he was telling us about it because it was a wild one. Well, the next morning, I was flying with Captain John Lucas—a great guy, who went back for a second tour in 1970 and shot down a MiG. He and I were going on a preplanned mission—to a truck-crossing or bridge or something. For that we had hard bombs on our airplane—fragmentary bombs. Major Fisher was on alert at Pleiku. Jump Myers and a guy named Hubie King were on alert over at Qui Nhon, which was over on the coast. Anyway, we were headed northbound when word came out that the camp had not fallen —it was still hanging on. The weather appeared that it was maybe getting good enough that they could get in some air strikes. They scrambled the birds on alert at Pleiku, scrambled the two that were on alert at Quin Nhon. We were contacted by radio and given what they call a divert—for "get where you are going." They gave us the coordinates of the camp at Ashau, and we diverted that way. The six of us all arrived in that area at about the same time. Of course, Fisher had been there before. He said, "Why don't we split up into flights of two and look for a hole in the clouds?"

It was complete overcast. We could not see anything, just clouds. He said, "If you find a hole, all the valleys run north and south. If you find one, we'll just punch down through thc weather and see if that's the right valley. If it is, tell us where you found it—we'll all gather and come back." We did split up.

Just to comment on the bravery of the pilots over there. If we could see the ground down through the clouds and had a good idea, we had a good chance of not busting our butts in the Skyraider. I can remember in one valley we went into that day—Luke and I—it was really tight for a Skyraider. Suddenly, here comes a flight of F-4s down the valley, looking, trying to get in to help that camp. Now, these guys' butts were hanging out a mile, because they're screaming down through there at 350 or 400 knots, and they didn't have much time or maneuverability. But they wanted to get in and help the camp.

Finally, we hear Fisher say, "I've found the valley." I don't know how you feel about faith or the good Lord, but the weather opened exactly over the place where you could have been sure where you were. There was a little dirt runway that was about three miles up the valley of A Shau, and that's where the clouds opened. We looked and there was the runway. Fisher and a guy named Paco Vazquez, his wingman, and Jump Myers and Hubie King, his wingman—the four of them got there first. They went down through this hole and preceded south down this valley to the camp that was under attack.

Luke and I got there and realized there wasn't room for six airplanes down in the valley—very tight quarters. Somebody said one time it was like flying inside Yankee Stadium, and all the cheap seats had the guns. We decided to hold up north in the valley over this runway and then, when these other fighters came out, we were going to go down. We just got down there, made about one orbit, and we got the call that two of the airplanes had been hit. The North Vietnamese had put their antiaircraft sights up on a hill, north of this camp, and they were actually shooting down on us, rather than up—as you ran the gauntlet. You go down this valley, and the firing is coming off these hillsides at you.

*It's easier to shoot down than up, isn't it?*

Well, I don't know. We were a lot closer. They assumed I think, that the weather would be good. Fisher said that Myers had been shot down. We didn't know where he was. We heard Fisher get on the radio, and he was talking to airborne command post. He said, "We got a pilot down, we need a helicopter to pick him up." About that time Lucas says, "We will come on down, Bernie, and we will cover you." So we start down the valley and as we're coming down, Jump Myers' wingman, Hubie King, was coming out of the valley and had his windshield blown away. He was in the open air now. He's coming out—he can't see anything. With Myers shot down and Hubie out of there, now there are four of us still flying. We get down there and Fisher says, "How long will it be before the helicopter gets here?" And the airborne command post said, "About twenty minutes." He says, "The pilot won't live that long. I'm going to land and pick him up."

The runway was made of pierced steel planking and when Myers landed on this "enemy hell strip," I heard Fisher say, "Raise your gear." Jump snatches up his gear so the plane collapses and is sliding on its belly. He had a 300-gallon fuel tank on the belly that burst when he landed on it. Of course, the airplane exploded in a fireball. Fisher thought Myers was dead, but then he saw him jump out of the fireball, run down one wing, and jump into a ditch. Again, a miracle—or whatever. Then, when the plane exploded, the fire went the other way, and the pilot escaped. Still, when Fisher saw all this smoke coming up out of Myers' flight suit, he thought the guy was burned. That's when he said, "He won't live that long. I'm going to land and pick him up." Fisher says, "I'm going to land. You guys cover for me." Lucas, I, and Bernie's wingman, Paco. There are three of us flying now. I thought if we drop our bombs, they're going to hit us. They would bounce right up and get us—they're very nondiscriminatory. The fragments would come up higher than we were flying.

The Skyraider had four 20-mm cannons and 800 rounds of ammo; that's all we had to shoot with. Very seldom, though, would you fire it all out. The gun would usually jam, for one thing.

We got in a chain, in a string, where we were trying to keep fire on both sides of the runway to keep guys away from Fisher when he landed. Fisher landed to the south because he thought that's were Myers was, but when he came down, there was so much fire from the airplane that was burning and the buildings on fire there, that when he broke out of the smoke he was too far down the runway to land, so he took off again, reversed his course, and landed the other direction. Then he had to reverse his course. I can remember seeing Jump get out of the ditch and start running for the airplane. Fisher had just stopped, when he set his parking brake, and was going to get out and look for the pilot. Then he saw his beady, red eyes coming up the wing. Prior to that, when I saw Bernie stopping on the runway, and I saw Jump Myers get up and run, I thought, "I don't know who you are, but you had better hurry." It looked to me like he was half loping, but Jump said he was running as fast as a forty-six-year-old man could.

It's comical now, but he we were listening to the guys inside the camp. And we got a call from one of them. He says, "Look, I think I am the last survivor inside the camp." As it turned out, he wasn't. The men were all isolated and hiding. He says, "I think I am the only guy left in here. You are cleared to hit anything inside the wire, or outside the wire—except the bunker in the northeast corner. That's the one I'm hiding in—other than that, have at 'em." We were just tearing things up on both sides of the runway—as best we could—inside the wire. I have an actual tape of the mission and on it, you can hear the ground fire and the cussing and the swearing and the whole thing. This kid on the ground says, "You can hit anything." Lucas, who was leading this thing, says, "All right, guys. Let's hit the camp." So we make a rip down through there. And the next thing we hear, the kid comes up on the radio. He said. "I know you guys are busy. I don't want to distract you, but that bunker you just hit is the one I'm hiding in." There was no panic in his voice. It was like he was ordering groceries. That's the type of reaction to the intense situation that was going on.

Then, Myers jumps in with Fisher, and we strafe both sides, trying to keep the bad guys away. It almost looks like we rehearsed it. Fisher taxis down, turns around, takes off. Just about the time he takes off, Lucas was on his wing. About the time he got his gear up, I was in the middle of the daisy chain—I was on his wing. And, as he went over to the far end of the runway, his original wingman, Vazquez, slid in on his wing, and away we went.

My flight lead, John Lucas, had been hit. He said, "I'm on fire." I said, "Luke, head for that runway up the valley." The helicopter that was coming to pick up Myers was just about there. And I said, "We'll get the helicopter to pick you up." He got the fire out and got on my wing. Lucas had been hit while this operation had been going on—he had no flight instruments, didn't know how high he was or how fast he was going. He had no hydraulic system left—he was just forcing the airplane around, but he stayed and pressed the attack.

One of the things you can hear on the tape was that we had no ammunition. On the last three or four passes, we were just bluffing. On the tape you can hear it where Lucas says, "Let's hit the fort." And I said "Luke, I'm Winchester." Winchester was a call for out of ammunition. He says, "I am too, but they don't know it. Let's keep making passes." So we made about three dry passes just to try to keep their heads down.

*You guys took lots of hits, didn't you?*

Yes, we did. It was the bravest thing I've ever seen anybody do. Of the six airplanes that got there, one was shot down—that was Myers; one had his windscreen blown away so he had to leave. I don't know how many hits he took. One had nineteen holes in it; one had, I think, twenty-three; one had thirty-one holes in it. And I didn't get hit. I was right in the middle of the chain, and I think the reason I didn't pick up a hit was because with three of us, we did not have the capability of always keeping fire on them. If you've got four airplanes, normally, you can always keep fire on them so they can't stand up and shoot at you. Lucas was ahead of me, so he protected me and kept their heads down as I pulled in. And I could protect his pull-off because I was behind him. And then Paco, who was behind me, could protect my pull-off I was the only one that had protection both rolling in and pulling off because Luke could not get back around in time to protect Paco as he pulled off. There was that much gap in the rope. They could shoot at Paco as he pulled off, they could swing their guns on Luke as he pulled in. I was very fortunate. They accused me of being one ridgeline over and not really being there. (Laughs)

As it turned out, it was the bravest thing I've ever seen: Lucas, what a brave guy. He stays there, his airplane on fire, had no flight instruments. Fisher, landing and picking a guy up on an enemy-held airstrip.

*Amazing!*

I have been asked why they didn't shoot Fisher. Why did they let the airplane land and pick up a guy? I fairly believe that they had thought that they had shot him down, too. I think they thought that he had made a forced landing there, and that they would finish him off at their leisure, that he was no longer a factor, and that they could run him down on foot.

*You guys helped by keeping them busy.*

We kept them so they couldn't get at him right away. They were directing their fire at us, I believe. In fact, Jump Myers, the guy on the ground, hiding in a ditch said the roar of the firepower being directed at the planes that were still flying was incredible. I'll tell you how close we were to him. The 20-mm shells are held together with a mechanical linkage—you've probably seen those World War II movies where they were spitting the stuff out. We were so close our linkage was falling on Jump, where he was hiding in the ditch. That's how close. It was very, very close. The last few passes we made, when we didn't have any ammo, I suspect we were as low as you could be—I don't know, twenty feet, fifty feet, thirty feet—I don't know. All you were trying to do was keep from hitting the brush and the very small trees. Just as close to them as you can—just to keep their heads down.

*That's "close air support" all right.*

Yeah, that it was. Close air support missions had to be the most rewarding missions that a pilot could have. You felt you were saving a life and doing some good—something by which you could measure.

*Do you remember back to when you found out for the first time that a pilot friend was dead? What did you think?*

Oh, we were angry. We were angry. I guess the thing that I remember the most was—I was having a beer with the Catholic chaplain, on an afternoon. Obviously, I was through flying for the day—if I was having a beer. Anyway, they called and asked him to come down to give the last rites to some Army wounded they were bringing in. And he asked me if I'd go with him. We went down to the hospital where they were bringing these kids in, and they were dead, not wounded. By the time they got there, there was no chance of saving them—they were already zipped up in the bag.

We went out to this shed behind the hospital, found the bodies—unzipped those bags. The chaplain went ahead and still gave them the rites of the church. You could see these kids had been ambushed. How the medics tried to save them. There was one that sticks in my mind. His chest was packed—they had packed it with all that they could. And there were four of them—all nineteen years old.

There was a twelve-hour difference between the west coast of the U.S. and Vietnam—it was maybe 2:30 in the afternoon when we went down there. And I was thinking—I was so angry—here were some nineteen-year-old kids and either their mom or their wife didn't even know they were dead yet. In the morning they were going to get a phone call—or a wire—that nobody wanted to get. The point is that I'm glad it happened at the end of my tour because it made you so angry.

The dead pilot brought the war to me personally. Before, most pilots felt, "I'll just fly my missions—I'm just doing what I'm supposed to do." It was not personal, not emotional. That one made it emotional. Really, I was angry. I wanted to hurt people. The reason I was glad it came late was because you can't fly like that, angry, because you start exposing yourself and exposing your wingman to things.

I mentioned the airborne command post that diverted us to Ashau. You could, on occasion, almost talk them into diverting you somewhere. When you checked in with them, you'd say. "Have you got anything going?" Usually, they did and you could say, "Gee, could you use some Skyraiders." And they would say, "Well, yeah." You could get them to do that in some instances. They would look to see where you were going and the importance of that mission. If you were to go out and destroy a wooden bridge or something, and they felt what they had was a higher priority mission—you could almost convince them into diverting you. I found that I was getting—I can't remember anything that came close to killing me—

emotional, you know, looking for trouble—if you will. You're taking yourself and your wingman in and just increasing the odds. The dead pilot, those four kids zipped up in bags—those I remember. Anger. My God, what a terrible thing.

*What was the key to learning to fly in Vietnam?*

Things happen so fast. It's just keeping up with what the hell's going on. There might be three different radios going at one time. Your forward air controller's talking to you on the radio. Then the airborne command post is coming in. Things are happening—click, click, click . . . ping, ping, ping. You're trying to figure out who's talking. Are they talking to me? What are they saying? It was combat. This was for real. You had to forget a lot of the things you thought were true from the movies—the John Wayne movies. That wasn't the way it was. You had to pay close attention, and you had to do it right. The first few months you're there, it's dangerous for you because you don't know what the hell's going on. Then, the next phase comes along where complacency can creep in. You think, "I've been here five or six months, and I haven't been hit, and I'm getting pretty good at this." So you'd start varying the way you'd break off the target. Guys would start doing some dumb stuff. I think toward the end of the tour, a lot of guys started thinking about getting home, which affected them.

*What about the Francis Ford Coppola depiction?*

I have to qualify that, because I was there early in the war. But I swear to God, that if someone told me that anybody in our unit was doing drugs, I would have said. "You're nuts." And I'd still tell you that. I renewed my faith in my fellow American because I saw acts of bravery on a day-to-day basis. And mostly to save lives. One guy, who as later killed over there, refused to hit a target one time because the bad guys had captured a village. They put all the villagers in the middle. The villages were built in little circles or squares, and they had the people—the innocents—standing in the middle of the village. The Vietcong were in the perimeter and firing out at what happened to be a South Vietnamese Army unit. They scrambled us over there to support them. Anyway, this Vietnamese forward air controller said, "Hit the village." The guys name was Bob Sanner. Sanner looked down and said, "I ain't gonna hit the village—those are civilians." He says, "I'll make a pass across there. If they shoot at me, then . . . but those are women and children." He made a pass through the village and says, "No, those are innocents—I'm going home." He picked up his flight and went home.

The helicopter pilot who exposed himself to multiple fire just because he knew the pilot wouldn't show himself until he knew the helicopter was there. To me, the Coppola thing was a mutation. I suspect some of that stuff happened over there, but the guy I saw there—it was a different thing. Brave. I was proud of them. There were times you were absolutely terrified. So scared to death. And anybody who said he wasn't was either not paying attention or was much braver than I was. I guess the thing that kept you going was, you knew, when I was there, that if you got in trouble, everybody—every flier, every soldier, every helicopter driver—would do whatever was necessary to save you—to pick you up. Like those F-4s that were trying to get into that valley. They had no business down there, but that's what needed to be done. I never heard a guy say, "Oh, I don't have enough fuel," or, "That looks pretty bad for me here," or, "I'm going too fast, I can't operate." Each guy says, "What can I do? What can I do?"

*Were there times that you just didn't want to go up?*

No. In fact, it was ironic. When you first got your orders to go over there, you thought, "All right, I'm going to do my job. No more, no less." I actually saw guys almost fight—get into fistfights—to fly. I was involved in arguments with guys saying, "Wait a minute. You flew twice yesterday, and I only flew once. And you're up there twice today while I'm only there once. I'm going to go in and get that changed." Really, guys wanted to do their job. They believed in it. They thought they were doing some good. Then, there were times—not that you didn't want to go up—that you just got scared.

*Remember any fun times?*

Oh, absolutely. We had some thirty pilots, and we had one jeep that was kind of our shuttle between up where we lived in the compound and operations down by the runway. Then we had one intelligence officer, a major, who had a jeep, too. So there were forty of us sharing one jeep, and this guy had one for himself. And, in the atmosphere, if you've ever been around fighter pilots, if his jeep was parked there, it was as good as yours, and they took it. He was always angry with us. His jeep was always missing.

Once, some poor enlisted guy wandered out and took this major's jeep. This major caught him. He was going to press charges on this enlisted guy, but we found out about it. This happened toward the end of the major's tour—he had three or four weeks before he was to go home. We said, "Don't press charges on that guy. The guy had just been drinking and had made a mistake in taking it. Don't prosecute the guy." But he says, "'I'm sick of this—that jeep always disappearing, and I don't have it when I'm supposed to have it." So a couple of us went to the base commander, who was a good guy. He says, "I think you're right. He's making a mountain out of a molehill. But I can't stop him from prosecuting this guy." We said, "Here's our plan. We'll go to the legal office—the judge advocate's office—here and get them to write up a phony paper that says, Yes, they agree that this thing should be pressed to the end—however, that he, the major, would have to be present when the court-martial was held." They also said that—due to the caseload—it could not be held for another seven months so that he had to stay for the court-martial. So we walked it on through. The base commander signed off on it; the judge advocate signed off on it, and it went through normal channels.

When the major got it, he absolutely went ballistic because when you get that close to going home, that's all you're thinking about. Anyway, he just went nuts. When he finally calmed down, we told him it was all a joke. He finally mellowed. "Yeah, I overreacted."

*Do you remember the worst day you flew in over there?*

Yes, the last month I was there we had forty-four inches of rain. The worst weather I flew in—we got scrambled off alert late one afternoon, almost at dark—and I went down south. We had an army company who had wandered into a North Vietnamese battalion command post. And it was getting dark when we got down there. There was a forward air controller that was airborne that had been supporting the operation and they were in a really tremendous firefight when we got there. Usually, when it got dark, we had a flare ship that would drop flares so we could see to strike the target. We got down there, and it was almost pitch-black, raining so hard. That night we had no flare ship.

I was leading the flight. My wingman, a guy named Kent Shatzer, got separated from me. But I found the FAC, who marked the target for me. I had napalm at the time on board. So he was directing my strikes against the North Vietnamese battalion. I was using his lights as reference. But I could not really see the ground. It was raining and dark. I would go as low as I could, down to the level of his lights, down to the position that he gave me, and I could drop. And, after I got one fire going, well, then at least I had some reference. Also, I was trying to get my wingman in with me—trying to talk to him on the radio. We had a device, an automatic direction finder, where—if I held my radio button down—he had a needle that would point at me. So I was trying to hold down the ADF, so Ken could find me, and trying to put this air strike in at the same time. The weather was really bad. We were down, I would guess, to maybe a couple hundred feet. The rain kept pouring. Ken said he thought he had me. I said, "Okay, I'll turn on my lights." I had my lights off so they couldn't see me. I turn on my landing lights, and the ground just exploded with ground fire. I turned off my lights and said, "That won't work."

Anyway, he did find me about the time I dropped my last can of napalm. He says, "All right. Where's the target?" I said, "I'll strafe in front of you. Ken, you follow me and you'll find it." I strafed ahead of him, and he put his ordnance on the target. That was the worst weather—night, no flare ship. Real difficult situation.

The forward air controller was absolutely the bravest guy, too. He was right down there, going over the target low and slow. We, at least, had some armament around us. We had guns. That did some good. They scrambled another flight after we got done. I believe they got a flare ship with them this time. Finally, it broke off the attack and the U.S. troops were able to go in and pick up their wounded and get them out of there. We got a nice note from the Army.

*Where would the war effort have been without the Sandys?*

Obviously, they would not have been able to get the rescue operations. Of course, there were hundreds of stories on individual rescue. You know, the PJs, the guys that came out of the helicopters, would go down on the ground and pick up the pilot if he was unable to get up there. And the Jolly Green Giant pilots would take those big helicopters in, too. I think I would be safe in saying that there would not have been so many successful rescues if the Skyraiders had not been available—just simply because of the capability of the airplane. And also, in close air support, I think that the army would not have had the confidence in calling in air strikes as close as they did. And that's just physics. There's nothing wrong with the jet pilot—it's just that he's going through there so damn fast. You blink an eye, and you've missed the target. We were, of course, going through slower.

*Where have all the Sandys gone?*

I don't think there are any flying at all. The last one I saw was at the museum at Wright-Patterson Air Force Base. In fact, its tail number is 649, and 649 happened to be one Bernie Fisher was flying. I think they've all been retired.

*Kind of sad, isn't it?*

Very sad. Great airplane. Great.

*Sort of sounds like we get a little too hung up on technology some of these times.*

Yes, we do. One of the things that guys were asked in interviews when they came back from Southeast Asia was, "What do we need for a follow on the aircraft to the A-1, the Skyraider?" One person said, "Some more A-1's. But give us an ejection seat"—which they finally did. "Give us some more reliable guns" was also a comment. It might not have been the guns, but it might have been the ammo. "Clean the airplane up a little bit." It had a lot of crap that was hanging on it that would slow it down. "Give us another 25 or 30 knots."

There was a Skyraider pilot over that, in a Sandy as a matter of fact, who had about six feet of the wing blown off. He was still able to fly it out of there. He was going to have to jump out, because the minute he let go of the stick, the airplane would fall off on one wing. He had to hold the stick way over because of one wing being six feet shorter than the other. So he takes the shoelace out of one of his boots, ties the stick over to the side, and jumps out. Then he's picked up. Incredible airplane.

I think the A-10 was the plane that was probably the follow-up plane. It's a heavy firepower, stay in close, tear things up pretty good. That's probably what they were thinking when they developed that.

*But the Skyraider sure would get you home.*

It would get you home. I've seen them land with one gear hanging, looking like a pheasant that has been shot. The landing gear wouldn't come down. But they'd get you home.

*Any practical jokes you pulled?*

In the KC-135, when you get flying up north—up over the pole—where the magnetic fields start to influence the compass, it is no longer accurate. There's a device in the aircraft where the navigator can actually take a little electric motor and drive the compass to whatever heading he wants. It completely lies to the compass. A motor drives it rather than it reacting to the magnetic pull. The compass that the navigator can impact with it is the one that the autopilot runs off of. So one night we were flying with some ROTC cadets, and one of the cadets was a bit pompous—I guess. He was ordering the other cadets around and things like that. So our boom operator came up to me—he felt that this young man could stand a little humility. We asked the young man if he'd ever seen a voice-controlled autopilot. "No, he hadn't," he said, but he really would like to.

We said, "Well, the way it works is each person has a voiceprint that is as unique as a fingerprint. We just insert a card in this computer and then the computer will pick up your voiceprint. Once it is accepted, it then will respond it will respond to voice commands from you." So we asked the kid if he was ready and we said, A-OK, we are ready with the card. And the navigator who was part of this ruse, said, Okay, go ahead. Tell us a few things about yourself. Well, he was so-and-so and he went to this school. He liked hunting and fishing, and all those things. "Did you get a good print?" he asked. "No, we didn't, you'll have to say a few more words." So we had the young man speak again. He was speaking right into the generator panel on the aircraft—it had some gauges and switches on it. That time we said, "We got a good print on you. Now all you have to do is tell the airplane where you want it to go. If you want it to turn left, it'll turn left; if you want it to turn right, it'll turn right." What we did was say to him to give the airplane a command. The young man said, "All right, airplane, turn left." And the navigator, with this little knob he has back in his panel, skews the compass around. Of course, the airplane doesn't do a thing. It just follows the compass heading that it's supposed to, so the airplane turns left and, of course, the young man thinks it's responding to his voice. So we have him do this—he has the airplane take two or three turns, and he is absolutely amazed. He has seen the state-of-the-art voice autopilot. We said now, "For heaven's sake, when you go back and sit down, don't say things like Boy, I'm so tired when I go home that I'm just gonna crash—or anything like that. You're going to have to be awfully careful what you say in the airplane. We explained that even though we had the card out of the computer, it can still be a problem. He went back there, absolutely delighted, but he was also careful what he said. That was our voice-activated autopilot. We had him hooked up good. Big-time! (Laughs)

*Any others?*

There was another story. I wasn't the author of it, though. We had a young pilot who's got a great sense of humor. It's his story. On a trip to Europe, they had a new boom operator on board. The young pilot told this young boom operator that he was probably aware of the global warming, and that it was a requirement that whenever we flew up in the higher altitudes—up over the pole—that when we went over Iceland, we were supposed to take an air sample. So he gave the boom operator one of those large, green plastic bags—just a normal garbage bag—and had him run up and down the aisle of the aircraft filling it with air. So he inflated it, and then taped it shut and put a lot of official information on it—like outside air temperature, altitude, the true airspeed, the Greenwich time, and the latitude and longitude of the air sample. He then had him turn it into base weather in Germany. Of course, they had no idea what the guy was doing.

The same pilot did another one that was good. We had a boom operator who'd been around long enough to know. But he's a good boom operator and normally you have to get up early in the morning to get ahead of him. Anyway, one day they were out pre-flighting an aircraft, and this young pilot asked this boom operator if he would take this snow shovel and go out and stand in front of the plane. They were going to tune the radar. So he had this guy with the snow shovel, walking from left to right, standing on this tiptoes out there. He had him getting down lower and lower an higher and higher with this huge shovel and, of course, everybody on the flight line was watching this big operation. He had this hook set in him big-time, too. It seems like we pick on new people, but everybody gets a day in the barrel. If they aren't messing with you, they don't like you.

Once we were coming back from Hawaii. You know, from the west of the mainland to the Hawaiian Islands is the largest stretch of water in the world. There's not a rock out there—nothing. And traditionally, when we're coming back, we allow one person to come up and talk to us and look out the window and see what we were doing. And we had a young female army reservist that came up and wanted to know where we were. We had a navigator who had picked up on that, and he explained to her how there were a series of buoys anchored to the bottom of the ocean —"Oh, every hundred miles apart." And these poor Coast Guard people had to live on these six months at a time, alone. They just bobbed around on these buoys that were 40 feet across. They were all there by themselves. And the Russian submarines would surface and antagonize them, etc. Of course, that's a total lie. There's nothing out there, at all. And we told her we were coming up on one of these buoys, and we had to give a the position report. We asked her, "Would you like to give position report because these guys seldom hear a female voice out here." So we wrote down a phony position report, and we switched the switch on the airplane so she was only talking interphone—she was talking to us in the cockpit. We had her read this, and she read a long position report off. And the navigator who was behind her and she could not see, responded. Boy, it was sure good to hear a female voice. He asked her where she was from, and she told him. Usually, when we do that to someone, afterwards we try to tell them that what we were doing was playing a joke. But she got away before we had a chance. So, someone in the world is an army reservist who thinks that there are a series of buoys between the West Coast and the

Hawaiian Islands, and the pilots fly from one buoy to another, westbound (Laughs).

*For all this craziness, pilots will close ranks if they are really challenge*

Exactly. They really care about one another. It's one of the few professions in the world where your lives depend on one another.

# Training at Farragut and Other Tales

## Gladys Buroker

Gladys Buroker has flown planes, taught students, both military and civilian, and lived a glorious life of flight. She described that life during a visit my ten-year-old son, Sayer, and I made to her Athol, Idaho, home, set back in the woods from the Silverwood Theme Park. As Sayer moved stealthily as a cat about the room, sometimes sneaking photographs with his Olympus 35mm camera, we proceeded with the interview. Gladys seemed delighted to see the boy's newfound enthusiasm for photography. And I would soon learn of the enthusiasms of a woman who stands out as a Northwest flying legend.

Over the years, Gladys logged more than 19,600 hours, eighty-percent of it instructing. But the breadth of her experience is what is remarkable—as I would soon learn.

The year was 1932, when eighteen-year-old Gladys took her first flying lesson. I ask her if she remembers her first flying lesson, and the interview is off and flying.

*She explains:*

I remember it very well. It was in an OX5 WACO 10. My instructor took me to the airport, which was in a cow pasture. He checked the oil but didn't check the gas—I guess he assumed there was plenty. He put me in the back cockpit and decided I needed a cushion to see out, so he got a cushion and showed me how to fasten my seat belt. He gave me a helmet that had two tubes, one on each side at ear level that came down to a "Y" under my chin. The single tube went down into the cockpit and up under the front seat. When he got ready to start the engine, he told me to hold the stick back. Of course, there were no brakes so you couldn't hold brakes. He went around and pulled the prop through and the engine started. He got into the front cockpit, put his helmet on, then pulled the tube from my helmet that had a funnel shape at the end around his neck. This is called a Goss port. He could talk to me, but I couldn't talk back.

*It's no fun if you can't talk back.*

There were times I wish that I could. He told me to keep my feet on the rudder and my hands on the stick and throttle. I felt him open the throttle, and we started taxiing down the field. When we got to the end, I noticed he gave a real blast on the throttle, and we swung around. I learned afterwards that's the way you have to do it when you have a tail-skid and no brakes. All aircraft today have tail wheels. He said, "We're going to take off," and I felt him open the throttle, wide-open, and I felt the tail come up. I was on the controls, and I could feel him on them too. We were off the ground pretty fast and in the air. He said, "Using right stick and right rudder, make a right turn." I did that, and I thought, "Gee, this is neat," and then I was making left turns. After several turns, he said, "Hold your altitude." Well, I hadn't even looked at the altimeter. I knew to hold my altitude I had to keep the airplane from going up or down. From then on it was just like riding over the waves in a boat, trying to keep it level. Anyway, before we'd hardly gotten started, it seemed that the lesson was over.

"Okay," he says, "it's time to go back. I'll take it." So I thought to myself, "I must have been doing some of the flying." Anyway, after he said that, I could feel the airplane diving, and it was just an instant or two before my head went down and almost hit my knees. He was pulling back on the stick and we went up into a loop. After that, the next thing I knew I was upside down. I found out later that he did what was called a slow roll. I remember saying, "Boy, that's enough. I'm getting sick." He kept it up and I really did get sick. I pulled myself up so I could get my head out of the cockpit. The next thing I remember him saying was, "Well, we're going to land--follow through." By then, I could have cared less what we landed on.

We landed, and I never did tell him I was sick, but I went and cleaned up the side of the airplane. I was afraid if I told him, he wouldn't let me fly again. During the next lessons, he continued to have his fun. I got smart and didn't eat on the days I was scheduled to fly. Once I started on takeoffs and landings, I was never sick again.

*That's great. What did your instructor say when you first went to him and said you wanted to fly?*

He just ignored me. In fact, I was working in a little combination grocery store and service station, and he came in for gas. I waited on him. I didn't know who he was, but after he'd left, the lad I worked for told me he was the pilot from out at the field. So I was watching for him the next time he came. in. I tried to talk flying with him. Told him I wanted to learn to fly, but he was more interested in visiting with the gal who owned the store. I did find out what it cost for lessons. It was going to cost twenty dollars an hour. I had five dollars saved up the next time he came in, so I asked him if I could get 15- minute lessons for five dollars. And he said, "Well, I guess so," but he still ignored me.

My boss was a friend of his, so the next time he came in, I got her to go with me to find out if I could get a lesson. Of course, I had no transportation, but at that time he was teaching high school in the morning and then did his flying in the afternoon and evening. She suggested that he stop by and pick me up on the way out and give me a lesson, then bring me back when he was through. That's the way it turned out. Still, he didn't think I was serious.

*Do you remember the point at which he realized that you were serious?*

I think so. I don't think I'd taken many lessons until he realized that I wanted to learn to fly and was working hard at it.

*Why did you want to learn to fly?*

I was born a girl, but with all the desires of a boy, shall I put it that way? I liked everything that boys liked—anything mechanical, anything with wheels. As I grew up, I had pictures of little scooters and cars I built, and models I made. That just seemed to be where my interest was, rather than in the house, cooking and sewing.

*You were really a pioneer of sorts, weren't you?*

That's what they tell me now. I didn't think much about it back then. There were no other women flying where I was. I was the first woman to fly in Whatcom County, where I soloed. That was in Bellingham, Washington. However, there were a few other women throughout the country that were flying. But there were not a lot. No.

*Did you get any catty remarks from men?*

No, if I did, I wasn't aware of it. I've had people ask me this many times, but I always felt equal to anyone in my same experience range, and was always treated that way. The fellows always treated me as an equal. The only thing that I can recall was, when I was training Navy cadets. I had just come in. There were places for the cadets to sit out in the sun, if they wanted. One day, several cadets were sitting just outside my window, and I heard one say, "I understand there's a gal teaching here." Another cadet said, "Yeah, and it would be just my luck to get her, too, wouldn't it?" (laughs)

*What year was that?*

Our first Navy cadets came in 1942. Our first government training was in 1939 in Olympia. Due to military restrictions on the Pacific Coast, we moved to Coeur d'Alene, Idaho, in 1942.

*What was the hardest thing about training for you?*

We had a certain amount of time to put in, and the students had to be at a certain level at that time. There were students we had to eliminate that would have made good pilots if we could have just had them a little longer. I think that was the hardest part for me. But I never really felt training was difficult. I mean, I loved every minute of it.

*Talk about the earliest plane in which you flew.*

I thought they were great for the period of time that we were flying. They had large wheels. You could land out in the fields, and you didn't get into trouble like you do with the small wheels of today. They had tail-skids and no brakes, but it was all grass or sod fields that you were using. The skid just slowed you down so there was no real need for brakes. I feel it was easier to learn to fly back then, than it is today. The sod is much more forgiving on landing. The airplanes were slower. The only minuses I can think of were that the engines were not as dependable and, of course, there were minimal instruments. The plane I learned to fly on had an altimeter, water temperature gauge, oil pressure gauge and a magnetic compass. That was it. No airspeed indicator, so you truly learned to fly by feel. I am very fortunate to have learned the joy of flying in the early thirties and not today, where you have in the cockpit a maze of gauges and instruments. It has become a concentration of instruments rather than a feel for flying.

*Did that scare you?*

What you don't know (laugh), doesn't bother you. And others seemed to be learning.

*Beryl Markham talked about how lonely it was to fly by instruments and without radio. Did you ever feel lonely up there?*

No, I can't say I have. I've made quite a few cross-country trips. There was always enough going on to keep me interested. Maybe she was in the clouds and flying solely by instruments. I have an instrument rating but have not done a lot of instrument flying.

*How do you look back on some of those pioneer aviators?*

Lindbergh was my hero. I thought he was sharper than Amelia Earhart. She was a woman, and I'm not saying this in a derogatory sense. She was just never the pilot Lindbergh was.

*Ever get lost flying?*

Oh, yes. I guess I'm ashamed to say that, really. That was right here, leaving Coeur d'Alene on my way to Burley, Idaho. I had my own plane and had three nurses with me. We were going down to a convention. I planned to go direct to Boise, but when we got down around Grangeville, the weather was solid fog so I turned back and landed at Lewiston and took on more fuel. Checking the weather, I was told I could probably make it to Boise if I followed the Grande Ronde River.

I only had a World Air Chart with me. Actually, I should have used a sectional chart for the trip I was making. I thought I was doing all right. The farther south I went, the better the weather got. Somehow I missed Baker, Oregon, all together. I never did see it. I knew from the time lapsed that I had passed it. I didn't know where I was, and I couldn't locate my position on the chart. (Thank heavens the three nurses were all sleeping.) After checking my time and compass heading, I decided I had passed the Boise Valley. I took a compass heading that I figured would take me to the valley. I finally came out in Ontario, Oregon. I had about two gallons of gas when I got there. That taught me a lesson. Never say you're too old to learn.

*How old were you when that happened?*

I was in my sixties.

*Any forced landings?*

Yes, I've had several. I had two forced landings in a Tiger Moth. I flew one on take-off, and one when I lost a cylinder gasket while towing a glider. I got down all right. Another time, the crankshaft broke when I was out with a student. In addition, I once lost oil pressure while flying with a student. While teaching three Army cadets during the war on cross-country training, one tank ran dry, so the student that was flying switched tanks but nothing happened. The students said, "What do I do now?" I said we'd have to land and told them to tighten their seat belts. I took over and talked to them on the way down. I had picked out a stubble field and landed OK. There wasn't a sound from the students from the time the engine quit until we landed. It turned out to be a vapor lock and the engine started after it sat on the ground for about thirty minutes. I had several others also but have never damaged an airplane.

*Did you ever repair some of your planes?*

I have done a lot of rebuilding, not so much maintenance. When I first started flying, they had a Waco biplane that had been damaged, and they were rebuilding it. My first job was cleaning the rust off the longerons on the fuselage. When I finished, they let me build the ribs for the wings. That led to making the envelope to cover the wings. Of course, my husband was a pilot and mechanic, and I used to work with him on doping the wings. I've worked with my neighbor Walt Redfern, who builds replicas of WWI airplanes.

*Did you ever barnstorm?*

Yes, in the thirties. I went with Jim Galvin, whose company is still running out of Boeing Field. They took Eddie Brown along, who was a parachute rigger, to pack my chutes. Eddie fitted the harness on me and showed me where the rip was. He told me to get up on the airplane and practice getting in and out with my chutes on. I would be jumping from a Swallow, which was an open cockpit bi-plane with a Hisso engine. I would be sitting in the front cockpit and was supposed to get out and stand close to the fuselage, and hang on until it was time to jump. Then I talked to the pilot. He said, "When I get on the jump run, I'll shake the stick." That meant I was supposed to get out on the wing.

Back then, they thought they had to pull the airplane up into a stall. You jumped when the airplane stalled. That way, you wouldn't hit the tail when you jumped. I practiced getting in and out of the cockpit with my chutes on before my first jump.

My chute was a backpack with the ripcord on the left side, so I would have to grab it with my right hand. While practicing, I found I could reach it fine, so I thought, "This is a piece of cake." Next day, when it came time to go, I can tell you the feeling is different. We were at 3,000 feet when I felt the stick wiggling so I got out. I can tell you that standing on the wing on the ground and standing on the wing going about eighty-miles-per-hour is a little different. I was hanging on for all I was worth. I'd let go real quick and try to reach the ripcord and, of course, I couldn't. I thought to myself, "Should I get back in and face the crowd a coward—or shall I kill myself?" (laughs) I decided I don't want to go back and face the crowd. When Jim pulled up into the stall and told me to jump, I jumped. I can tell you, I wasn't long in getting hold of that ripcord once I left the airplane.

*Did you do a lot of barnstorming?*

I had agreed to make twenty jumps on that trip. I was injured on the seventeenth jump, so that was the end of my jumping. I bought my own airplane after that, a Travelaire 2000, which was an open-cockpit, three-place biplane, and barnstormed the Olympic Peninsula. That was before I had a license to carry passengers.

*What year was that?*

That was 1935. I actually had enough hours of flying to get a private license, but somehow it didn't seem important to me. Anyway, I had a pilot friend who went with me. We liked to go to Forks as that was a good spot. It seemed like the people there—the loggers and the Indians—always had money. As we approached Forks, I'd get out and walk along the front spar, hanging onto the flying wires. Near the end of the wing, I could hang on to the wing strut—it's a little more stable, easier to hang on to, at least. When I was out there, the pilot would come in low over town—or wherever he saw people—and I'd wave. By the time we landed, we would always have people there ready to go for a ride.

*That was how you made your money—on rides?*

That was the way it was parachuting, too. When you got them out, you'd sell them rides.

*What did a ride cost?*

I can't remember what they charged when I was parachute jumping, but with my airplane, I was charging five dollars a flight for two people, so it would be two and a half apiece. When I bought the Travelaire, it cost $400. I had three hundred in my savings account and had to borrow the other hundred. I remember I was able to pay off the hundred in nothing flat.

*What exactly happened when you got injured? How did it happen?*

At the time I got hurt, we had two airplanes. Frank Kammer of Wenatchee had joined us with his Curtis Thrush, a four-place cabin monoplane. I would alternate my jumps between the Thrush and the Swallow. The day I got hurt, I jumped from the Thrush, and it was probably the best I'd done landing by the crowd. I could see I was going to land on the strip we were using. When I was about a hundred feet from the ground, Jim took off under me in the Swallow. I have never known for sure, but I feel that the turbulence from the Swallow partially collapsed my chute. I know I hit the ground much harder than ever before. I ruptured the cartilage in my right knee.

Once I was assisting on the rebuilding of the plane I told you about, and was asked if I'd like to fly to Boeing Field and pick up rib tape and cord they need. Everett was as far south as I had been--I had never been to Seattle. I said "Sure," if he thought I could do it, I was game. I'd never been out of a five-mile radius of the field. "How will I get there—I've never been there?" He just went over and got a road map and handed it to me. I collected my gloves, helmet, goggles, and jacket and went out and got into the airplane before he changed his mind. I asked about fuel. His reply, "There's plenty." When I got out of the five-mile radius, things were not familiar.

I had lived in Bellingham all my life, so I knew the highway south. I decided to take a look at the map. I took my gloves off and sat on them. I held the map with one hand, and the wind vibrated the map so I couldn't see a thing. I held the stick between my knees so I'd have two hands, but it was useless. The wind coming through the front cockpit made it impossible to read. I remember thinking at the time, "He just thinks I can fly by the seat of my pants, but how do you navigate by the seat of your pants?" I remember thinking that. (laughs) I knew the highway went to Seattle and I was sure I could make it to Everett. I had been there by car. I soon became apprehensive. I couldn't recognize anything, but still I could see what I thought was the highway, which was the only paved road at the time. I finally saw smoke up ahead. Then I knew it was the mills of Everett. The highway was easier to follow once I was past Everett, and I did make it to Seattle and Boeing Field.

*What was your next most ambitious trip?*

By '36, I had owned two motorcycles. The first one was a four-cylinder Indian Ace, which I traded in for a new Harley 45. I had a girl friend, and we thought we should see more of the country than northwest Washington, so we decided to take a trip and see every state in the United States. We both worked and saved until we had a $150 apiece. We left the first of September 1936 and were gone three months. We visited every state, up into Canada and as far south as Mexico City.

*That must have been unusual for the times.*

Well, it was, in fact. Very few people we saw realized we were girls. We slept in sleeping bags and always filled our gas tank in the morning at good-looking service stations so we could clean up.

One morning we pulled into this station, and my friend went into the rest room. The fellow waiting on us, a middle-aged fellow, started laughing. He said, "You know, I thought she was a fellow, too." So I just laughed with him. After I paid him, I headed for the rest room, too. When we came back out, he was standing in the door shaking his head and laughing. When we got to New York, we wanted to see the Empire State Building, so we pulled up and backed into the curb. We got in the elevator —it must have held 25 people—and as we started up, I could see this fellow staring at me. I finally turned and looked at him. He said, "Are you girls or boys?" I said, "That's an insulting thing to ask a fellow." Everyone in the elevator just roared.

*What else happened?*

In Mexico City, before they had traffic lights, the police directed traffic on the corners. We were just leaving the city on our way back to the U.S. when a policeman pulled us over the side of the street. I told my friend, "If he speaks English, act like you don't understand him. I don't know what I've done." He talked Spanish, and we couldn't understand him. He finally started pulling at his ear. He didn't act mad or upset. People started gathering around and, after about ten minutes, I realized he was trying to find out if we were girls, so I said, "Si senoritas." Everyone laughed, and he sent us on our way.

*You flew planes across country too, didn't you?*

Yes, especially during the war. Most of the cross-country flying, to start with, was picking up aircraft that had been assigned to us for training. We taught primary, secondary and cross-country courses. We'd take three students at a time for the cross-country course for a three-hour flight in a four-place cabin airplane. The student would change every hour, alternating with the flying, the radio work, or navigation.

*Any tough situations on those flights?*

I had a couple of forced landings. We were flying Fairchilds, which have two wing fuel tanks. When one tank went dry, it would form a vapor lock, and the gas would not run from the full tank. The first time this happened, I landed in a field in Missoula, Montana. As soon as I was on the ground and tried the engine, it started. I found transportation for the students to Missoula Airport. I was not wanting to have them with me for the take off. I flew to the airport and picked them up. We returned home. I wanted the engine checked before another flight. This happened many times, especially during hot weather. We finally got smart. Every time he had to switch tanks, we made sure we were over an airport.

*Of all the planes you've flown, what was your favorite?*

I'd have to say the Fairchild-24 with a 165-horsepower Warner. It was just such a smooth, comfortable airplane to fly. For fun flying, I'd choose any of the open cockpit Waco bi-planes.

*Were you good at judging the length of clearings?*

Yes, or maybe I was just lucky. You get pretty sharp after one or two forced landings. I think I've improved even in my old age, because the last twelve years I have done a lot of soaring and gliding. That makes you really conscious of wind direction and velocity. I've also had my own balloon and did quite a bit of ballooning. I think it's kind of what I said in the beginning. I always liked anything with a challenge.

*Any balloon stories?*

I think the time I laughed the most was right after I got my balloon license. I started to teach my son-in-law. On our second flight, we were heading south. About the time we got over the rimrocks, the wind really started picking up. I was really inexperienced—even though I had my license. I really didn't have enough experience to be teaching someone else. So I told Doc we'd have to land. We kept looking, trying to find a field that we would line up with, because with a balloon, you have no control except vertically. You just go where the wind takes you. We saw a field ahead where I planned to land. We were going so fast that it was going to take both of us to land. So I said, "I'll handle the burner, and when I tell you, pull the ripcord." That opens the top and lets the hot air out.

Just as we cleared the fence I yelled, "Pull the ripcord." I closed the burners and we hit the ground hard—throwing Dick out of the gondola. Here he is on all fours, trying to keep ahead of the burners. I mean it was funny--at the time anyway. The wind was so strong that it dragged the balloon, even though the apex of the balloon was wide open.

*Do you feel like you've learned anything about the winds from flying gliders?*

Yes, I'd say you become more conscious of the wind in a glider. Let's say your glider comes in on an approach for landing at 60 miles per hour. If you have a 10 mile-per-hour wind as you turn into the wind on your landing approach, you have to increase your speed to 70 miles per hour. in order to penetrate that wind and get to your runway. You do really have to be aware of the wind, and your correction that way. To me, gliding is the safest type of flying there is.

*Is there any most engaging image you remember from your flying in those earlier days?*

I think the thing that impressed me the most and has stayed with me the longest was my view of the earth while jumping—it's always a great sensation. You're just floating, and then I was always amazed at when you get down to about 100 feet above the ground, you don't have any sensation of falling at all. It just looks like you're standing there still and the earth is coming up to meet you. It's a different sensation.

I've had a few pretty things. Three times in my flying career, I've had a rainbow completely surround the aircraft. Not all pilots get to see that. It's being just in the right position between the clouds and the sun when atmospheric conditions are right that the rainbow is a complete circle. The last time that happened was a few years ago while giving a young girl a glider lesson. The clouds were about 2,000 feet, but there were holes in them. I had them tow us up through a hole, and we went out on top with the glider. When we released from the tow plane, we just happened to get in a spot where the sun was right on the moisture and the clouds. No matter where we turned or where we went the rainbow followed us. She was so excited about seeing it. She still talks about it when I see her.

I have never forgotten the day flying over the Cascade Mountains flying from Kennewick to Seattle. We were above the overcast, and I spotted something ahead of us. I noticed I was gradually gaining on it. When close enough, I could see it was an eagle, and it was just going for all it was worth looking back to see we were gaining on him.

*What were you flying?*

A Curtis Thrush. One time in '35, my husband-to-be, Herb, was going from Everett to Wenatchee to pick up an airplane. A young fellow that owned an airplane was flying Herb over to get some cross-country experience. The fellow's wife and I were taking the bus over to Wenatchee and would fly back with them. When we got to Wenatchee, there was no sign of them. We stayed around for awhile and finally got a call from Herb, saying they had cracked up in the Cascade Mountains, and we'd have to come back home. They were flying above the overcast. This was before airplanes had carburetors with heat.

It turned out the plane would lose power and let them down into the clouds, and then the engine would pick up power and they would lift up out of the clouds. That happened about three times. When the engine continued to lose power, they knew they were going down. The sensation in the clouds is that it stays pretty milky, but if you're close to the ground, it turns dark. When it turned dark, they came back on the stick, to slow the airplane down. Just about the time they came back on the stick, a big pine limb brushed by the airplane, and they pulled all the way back on the stick to stall it. One wheel hooked over a limb of a big tree and they just slid up against the trunk of the tree, stopping with the wheel hanging on one side. They were about 40 feet up in the tree. They had tools with them and were able to disconnect one of the control cables to help them get down from the tree. Neither of them got a scratch. As they were getting down from the mountain, they both got blisters on their toes.

*That's pretty remote country.*

Yes, for a number of years afterward, the CAA—Civil Aeronautics Authority—would get reports from hunters who saw that airplane up in the tree.

*What were the worst flight conditions you've flown in?*

Well, I've had a few things that kept me on my toes. One happened when we moved to Coeur d'Alene from the West Coast. All my past flying had been along the West Coast, and I had never flown in snow conditions. I'd never even been in the air when it snowed. In order to get airplanes over here, they had to be ferried, and it was decided that I would take the six Piper Cubs that we were using for primary training. I would fly one, with five student pilots following. None of the students had received cross-country training. We couldn't come direct because the pass was closed in--so we left Olympia, flew south to the Columbia River, then went east to The Dalles, in Oregon, where we refueled. The weather was minimal--this was in February 1942. As we headed east, my main concern was the responsibility of the students following. Before we got to Sprague, Washington, it had started to snow, and I was concerned. I had never flown in snow. There was an airport in Pasco, but when I did some figuring, I knew we would not have enough fuel to get back.

The conditions were getting worse, and there was no airport before Spokane. I knew we were going to have to land—and believe me—I had been looking. I was over Sprague Lake before I saw it—a field I figured would be all right. I made a circle to make sure I had all five airplanes behind me, and I did. I circled out around and came in and landed. I got out of the plane, stood on the ground and signaled to everyone as they'd come in, whether to land or go around again. The place had big cottonwoods that the students had to come in over. Eventually they all finally got down.

*How did you learn to deal with that stress?*

Well, I'm not sure that I did. That's one thing about instructing: there's certainly a lot of stress. I think I felt it all my life, but I do think it got easier with time. I've done enough instructing to where I feel I'm a better judge of what students can do or what they're ready to do. One day out here, about six or seven years ago, when we were flying passengers in the glider, an airline pilot and his son came out and wanted to go for a ride, so I took his son up. That was a two-place glider, one passenger and the pilot. When we got back, I took the dad up. I always put the passengers in the front seat, so they could see much better. Just after take-off, I could see big rolls of dust off to the west. I thought, "That sure doesn't look good." We go to 2,500 feet before we release, so we went right on up to 2,500 feet, and I could see that a storm was coming. When I released, the tow plane went back to the field. There's lots of lift in front of the storm, so I was watching the tow plane and saw him try to land. He was drifting clear off the runway, and I never did see what happened to him, but I knew that he never landed on the runway. I knew I had to land and could see that there was a direct crosswind. I started losing some altitude and told the captain, "We have a real strong cross-wind—a glider can't handle the cross-wind that we have and land on the runway, so I plan to come in between the hangar and the cafe and land cross-ways on the runway." He turned around and said, "You can't do that. Look at all those trees!"

There were a lot of trees, but I knew I'd never get to the trees if I planned correctly. Well, here's an airline pilot, and you know how long a runway they have to have. But the guy was upset, petrified. He sat there and never said another word. I said, "I don't really have another choice—it'll be all right." I wanted him to know what I was doing, so he might be less frightened. I was heading into the wind sideways, until I turned in for landing. It was summer, and there were a lot of people out there watching to see what was going to happen. Fortunately, we got down all right and didn't roll ten feet when some of the Aerodrome crew grabbed the wings.

*What was the main change over the years in flying?*

When I learned to fly, the air was free from the ground as high up as you wanted to go and horizontally the same. Today, you can hardly go ten miles without invading some type of airspace. Things are so controlled now that there just isn't any freedom any more. That's what I notice. I'm not saying it's bad, but I sure am glad I lived and flew when I did.

*Do you ever long for the good old days?*

No, not really. I think I've finally adapted to the change. It is kind of fun to tell people how it used to be. Sometimes, I think they think we're exaggerating, but I'm sure that aviation—the way I saw it back in the early days—was pretty much like it was for the pioneers, who came west on the wagon trains. Now they have highways, and so on. We had no airways. The first airways I remember were towers with beacons on them. They had them between Missoula and Spokane and clear to Seattle. About every 25 miles, they had one of these towers with a beacon on it. The pilots could fly at night and follow those beacons. I took up nursing in the '60s. I worked at the hospital. After work, nothing was more therapeutic than getting into the airplane and going flying. It was restful and I enjoyed that. But whenever I'm out flying, I'm thoroughly enjoying myself. I love to go out—especially with the glider. I do lazy eights and wing-overs and just kind of float around like a bird.

*Do you think you have some secret wish to be reincarnated as a bird?*

Maybe.

*Beryl Markham, again, talked about developing an intuitive sense—or a fatalistic approach—and she said that most pilots she knew were like that.*

I think most people who fly for any period of time have a fatalistic approach all right, but really never felt pessimistic. I feel that, if I got killed in an airplane, it would probably be a mid-air crash. I just don't feel like there's any reason for having a problem otherwise. Of course, there are circumstances. My brother was killed in an airplane. He was only 1,000 feet in the air when the horizontal stabilizer let go—and he wasn't high enough to parachute out. He tried to, but the airplane hit the ground before he could even get out. So there are things like that one cannot control.

*That must have been tough, losing your brother.*

Yes, it was, and especially so because I had taught him to fly. We were really close. Yes, it was hard.

*You don't get over things like that.*

No, you don't.

*But you continued to fly.*

Yes. There again, the thing that was hard for me was that my brother more or less idolized me as he was growing up. I was doing everything he wanted to do. And, as I said, I taught him. He was a fighter pilot in WWII. When he came back from WWII, he and Clay Henley, who were friends, built a midget racer. They went back to the Cleveland Air Races to fly it. My brother was doing the flying. They did so well that they decided to build another one and make it better—faster. They built this one with a cantilevered horizontal stabilizer. My husband, who was an aircraft mechanic as well as a pilot, told them they should put brace wires on the stabilizer as a safety precaution. They didn't want to do that because it would cut down the speed.

In the meantime, my brother told Clay he would help finish the airplane, but he didn't want to fly this one because his wife was expecting a baby and he just felt he shouldn't. That was fine with Clay—he said they would make the cockpit a little bigger and he would do the flying. Milton Chester, a CAA engineer, stopped by to see the airplane and ran some tests on the stabilizer and told them it was fine, so they didn't even consider wires after that. After the plane was finished, Clay still couldn't get into the cockpit. So my brother changed his mind and said he would fly it. The plane had been flown about eight hours and was making one more flight before heading to Cleveland. It was on this flight that he crashed.

*It's been a quite a life of flight.*

Yes, it has. Not many people can go through life earning a living at what they love to do and still have a nice family—I have three children and six grandchildren.

*Do you think you'll ever stop flying?*

I plan to fly as long as I can pass my yearly physical examination. I have a couple of airplanes down there in the hangar, and I sure don't want them to get covered with dust.

# Soaring

## William C. Sproull

Bill Sproull's dad was a WWI aviator, who flew in France. The legacy of his dad's flying no doubt held sway on young Bill, who went into the Navy V-5 program near the end of WWII. After two years at the University of Kansas, Bill entered active duty at Pensacola, getting his wings just before the Korean War started in June of 1950. He spent four years as a naval aviator with the Pacific Fleet in Korea and off the China coast before coming back to the U.S. to complete his Aeronautical Engineering and MBA at the University of Michigan. Following graduation, Bill went to work for Honeywell for 33 years, but the love of flying surfaced again when he became interested in sailplanes and soaring. The year was 1974 and for the near 25 years since then he has become one of the sport's most outspoken advocates. In 1997, he was elected to the National Soaring Hall of Fame at the National Soaring Museum in Elmira, New York.

*What got you interested in soaring?*

I was on a business trip in Phoenix and saw an ad for soaring demo rides, so I went out and talked to the operator. I gave him a little of my background. I told him that I'd really like an introduction to soaring rather than just a demo ride. We went up and stayed up for quite a while. It was just a great thrill. It was the first time I had been in lift where you had any ability to go anywhere. I came back to my home in Bloomington, Minnesota, and looked up the Minnesota Soaring Club, got my rating and that was the start.

*What impressed you from that first flight?*

The freedom and the closeness to nature. I'd had flown over 2,000 hours with propellers sitting out there, so the idea that you could stay up in the air and thoroughly work nature was something new. I loved the whole feeling of using my intelligence and knowledge to stay up. I think that was the profound thing. Of course, it wasn't long before I encountered a hawk, and once you get in a thermal with a hawk, you're really hooked. This occurred back in Minnesota. I was up in solo flight and looked up and there was a red-tailed hawk right off my wing tip. He looked over at me, and I looked over at him. I said, "Holy smokes! This is what it's all about." It was a wonderful feeling. I remember coming home and telling my wife about it. This was one of the most impressive things that ever happened to me. We were probably together for three or four minutes, then the hawk went off, but I rejoined him in another thermal. A great experience!

*So you learned a little bit about how birds fly.*

Absolutely. I found out he could turn inside me, and he could find the best lift far better than I could.

*If he was a warplane, you better look out.*

That's for sure.

*How long before you felt comfortable flying a sailplane?*

I think it took about a year or two. Soaring is a challenge in many disciplines. You have the energy management discipline, the aerodynamics, strategic and tactical decision-making, weather forecasting, and anticipation. You need to choose carefully the terrain you want to fly over and be able to set and achieve goals.

*Anything you had to unlearn from flying power planes.*

Any power pilot getting into soaring comes to realize that there are a lot of options up there. If a power airplane quits, you have very few. I approached soaring mainly from a technical standpoint, but I've known plenty of people, especially contest pilots that don't approach it that way. Often it's more of a judgmental type of thing.

*What do you mean judgmental?*

It has to do with the feel of flying. Feel is probably the biggest thing. When you're looking for lift, you're feeling it. You're looking for one wing going up as you try to center into a thermal. It's a matter of feeling what's happening to you.

*Tell me about a strange landing.*

One was in a fairly tall cornfield. The landing taught me I'd better make up my mind a little earlier before I get committed. The field I wanted to go into looked like it was not available. When I flew over the top of that corn, I realized I was going to land somewhere in that thing. I tried to pick a place fairly close to the road, so I wouldn't have to walk out too far. As I landed, I remember hearing the sound of corn contacting the horizontal stabilizer. Then all of a sudden, I fell into the corn and stopped quickly. I bet I didn't go any more than 30 or 40 feet. It was very much like a carrier landing.

*Explain?*

I guess my first carrier landing was very much like landing in the corn. At the time, I could see the LSO was giving me pretty good signals. Next, he gave me the cut and I dived for the deck. You are going so slow and then you cut back on the stick. For that first carrier flight, I probably pulled back too soon, floating down the carrier deck and hit a wire while still airborne. It suddenly pulled me out of the air. Sudden as the cornfield landing.

*What did your friends say when this fifty-something year old man took up soaring? Did they think this was middle-age angst? Did they tease you?*

I don't think they teased me, but you could tell they were wondering about me. "That's pretty dangerous," they'd say. And I think soaring gets that rap, but I became convinced very soon that this was not a dumb decision. Sure, you can make it dangerous by over flying terrain upon which we could not land. Still, I don't feel compelled to convince people that I'm not crazy. Most of them will give you that kind of reaction, though.

*Tell me about your first off-field landing.*

My toughest off-field landing was probably my first. I was actually in a contest. I got good lift up to cloud base then I made a run to the turn-point to take a photo, which is what we do. But when I returned to the thermal, I found it had died. That was the first time I was exposed to the life cycle of thermals. I was faced with a situation of deciding what to do as the flight was over. Once you've landed off-field a few times, you get the necessary confidence, so it gets to be more a question of determining the right field, the best landing direction, and the approach and landing techniques. But the first time you make a landing like that, you don't have the confidence and therefore spend extra time thinking about the wrong things.

*Biggest misconceptions about sailplane flying?*

The biggest misconception is that you don't have control over what you're doing. The second is that the only thing keeping you up there is the wind and that if the wind dies, you're going to be on the ground. The third misconception is that if the lift dies, you're going to crash—you have a very bad situation. All of these things are just not true.

*Talk about glide ratio. That's a key aspect.*

Glide ratio means the distance in feet you move forward for every foot you lose in altitude. Glide ratio is built into the ship that you fly. My ship, the LS-6, has a glide ratio of about 40-1. I've often flown 40-50 miles without losing any altitude by picking thermals and doing what we call "dolphin soaring." I think that sailplanes allow you to know that in every dimension you are free. It's such a joy to be up there.

*How strong are thermals and how long can they last?*

One time on a flight in Phoenix, I hit a thermal, and the variometer just pegged at over ten knots. It seemed like I went from 1,500 feet above sea level to about 12,000' in about five minutes. My biggest flight was about six and one-half hours.

*Since your dad was a flier, what did he say when you became one years ago?*

I think he was very proud, because I approached flying differently than he did. He was not really into "aviation." He had volunteered into the air service during WWI, and the more I talked to him, the more I began to understand that he was not as confident with airplanes as I might have thought he was. Aeronautical knowledge probably drove me to know airplanes from many disciplines. That's why I studied Aeronautics in school, engineering at Honeywell, and got into flight control. In my flight control work, I picked up a couple of patents in aircraft and spacecraft flight control. Specifically, the patents were for adaptive flight control and multi-mode reaction controls for spacecraft.

*Any personal things your father told you?*

Before he died, my father and I had a chance to fly together in both power-driven planes and sailplanes. The sailplane flying was an especially great experience. He was probably 75 at the time—it was different. The thrill he expressed when he talked about soaring after that flight told of a great experience. On that flight, we caught some thermals and generally had good lift. I flew around initially, but then he flew it for a while. The biggest thing I feel is that my father and I didn't do enough talking—on the emotional side about flying. We spent our time talking more about the mechanics and operational aspects of flight, instead of how he felt about it. Nonetheless, I think he was very proud.

*There seems to be something highly environmental about sailplanes.*

Yes, there is. You feel good about it. I certainly have driven around power planes polluting the atmosphere, but I feel awfully good about flying on my own. For example, I've flown in the Grand Tetons in power planes and would see hikers and would feel as if I was cheating on the world a little bit. On the other hand, when you've got your sailplane, and the other people up there have their hiking stuff on, you both feel you're doing the environmentally sound thing.

*What was one of your closest calls—in the military or in soaring?*

It was in the military in 1954 in the Pescador Islands, off Formosa. These are now known as the Taiwanese Islands. I took off in a seaplane with waves 30 to 40 feet high. Then the moment you put the power on the engines, the water spray is so heavy that it's absolutely impossible to see. We were on instruments for the entire take off.

Another situation occurred about midnight over northern Japan in 1951. We had failures in both engines. We were flying at 10,000 feet, over 6,000-foot mountains. (I was the copilot on that flight at the time, but was back at the navigation station.) When the second engine went out, I just yelled for everyone to leave. I got my parachute, but the pilot made one last effort to start an engine. Luckily, it started.

*You weren't indecisive. It seems like you talk about making decisions on sailplanes as well. Are you the kind of person who likes to be decisive?*

I think so. You like to know you're in control. But I don't consider myself a risk-taker.

*Think you'll ever stop flying?*

Yes, I think I will. I think every pilot has to make that decision. When that time comes, I hope I make the correct choice. Besides, I've got enough memories that will keep me—and I can always fly with others.

# Flying Dignitaries

## George Jobin

George Jobin was born in Detroit about a mile and a half from where his grandfather's farm had been. His grandfather had come from Germany about two years after the Civil War had ended. George attended public school in Detroit and then Wayne University, which is today Wayne State. He enlisted in the military in January of 1942, but didn't get called up at that time. A student at Wayne at the time, George went during semester break and took his physical, which placed his name on the waiting list. The reason for this was that at the time the military did not have adequate facilities to take people in the service as they came in. In the interim, he returned to school. Finally, on September 3 of 1942, George Jobin got called to active duty. As he explains in the interview, he wanted to fly in the war effort, but George's talent was sought elsewhere. A highly skilled pilot, the military selected him to fly government officials around in the States. Initially, George and another Ensign received an assignment to Anacostia. Two months later, two other Ensigns were assigned the same duty. George and the three others were the only Ensigns assigned to this duty during World War II.

Following the service, George finished school while working a forty-hour week at his job at General Motors. He worked in engine design there and then went to Ford Motor Company for about six years. He later served as plant manager for Sealed Power Corporation in Mexico, for three years and returned to work as original equipment sales manager for the same company in Detroit until he retired in 1994.

*Was that frustrating? Were you chafing at the bit?*

I had taken what they called a CPT course in pilot training for college students. I had taken one course and then a second course. Finally, I decided in the spring 1942 semester while I was still in college that I could get a third course in, so I took an aerobatics course. I was getting ready to go and summertime came along and I still hadn't been called, so then I took a course in celestial navigation and in meteorology at Wayne State.

*You sound like you were determined, doing a lot of work without any certainty of payoff aren't you?*

That's right, but I loved it. I figured that when it came time to go into the service I'd have no problem, and I did not. I zipped right through training in about two-thirds of the time it was taking some of the fellows who lacked some of the background I had.

*Was there any funny thing that they taught you in those early days, anything that you'd learn to be total nonsense later on?*

I can't think of anything that went on during flight training that I didn't feel was a necessity. We had fun, particularly with flying formations in single engine planes. We were flying a single engine Boeing, what we called the N2S. It looked very much like the Navy N3N. We went through primary flight training at Pasco, Washington, in December 1942. In September, we were sent to a pre-flight school at Iowa. All the cadets were being sent to what they called a preflight school. We'd get our military indoctrination and Navy history, etcetera, and ground school, preparing us to go into flight training without having the encumbrance of too much classroom work. When we finished in Iowa, there were twenty in the group. We were originally designated to go to the Navy flight school at Grosse Isle, Michigan. The Brits were also training there and had sent in two or three times the number as they were supposed to, so the U.S. cadets were sent out to Pasco, the twenty of us. We rode the train out.

*What was the funniest thing that happened in training? Did you have any close calls?*

I had no close calls. I only had one flight check down. When we were going through flight training, we would periodically take check rides with instructors and there was only once in primary when I didn't hit the landing spot I was supposed to hit and instead had to go back for a second check.

*That's a pretty good record, isn't it?*

Yes, I think this is why I ended up with the job I had after I finished training and got my commission in mid-June 1943.

*You ended up with the job you did have because of your quality flying?*

I think so.

*Explain what your job was.*

I was sent to Washington, D.C. where the military assigned me to the Naval Air Station at Anacostia. Nobody in flight training at Corpus Christi, Texas, knew where the hell that was. I asked the first guy at the cadet office and he said "the Aleutians." Another one said, "Well, I think it's up in Greenland somewhere." I said, "Let's look it up in the book," and found out it was Washington, D.C. I reported for duty and found out that I would be flying VIPs as a co-pilot of a twin-engine transport. I had gone through flight training where I ended up flying PBYs, so I had seaplane and twin-engine qualifications.

*Does that mean flying VIPs, meaning important persons?*

Yes, I flew the Secretary of the Navy, James Forrestal, the commanding admiral of the U.S. fleet Admiral King, Frank Know and Admiral Byrd. We also flew various other admirals, congressmen and senators.

*When I was a kid, I met Admiral Chester Nimitz. In fact, my dad was a good friend of his. I wondered if you ever flew him?*

No, he had his own plane and would come in from Pearl Harbor on what we called PB4Y, which is an Army B-24.

*Give me a favorite story about flying a VIP.*

One of them involved Admiral King. We'd been down to Norfolk. He'd gone down to check the Atlantic fleet out. On the way back at night, we were flying and he came over the intercom and said, "Turn on the goddamn lights. What do you think I am an owl?" About fifteen minutes later he said, "Turn off the goddamn lights. What do you think I am an owl?" He repeated himself. He was not one with much of a sense of humor. He lived on board the U.S.S. Dauntless, which was about a 250-foot yacht, which had belonged to one of the Dodge brothers of the automobile business. He had a gun mounted forward on that and left its berth at the yard. The story was that he came off the boat one day and told a sailor on the gangway to do something. The sailor said, "Aye, aye sir. Your every wish is my command." He came back and said, "You're goddamn right it is."

One of the funniest things that happened is that when we had taken James Forrestal, who was secretary of the Navy. We took him to his home in New York City, where he likely had some business to do. We stayed overnight and were to pick him up the next morning at La Guardia airport to fly him back to D.C. We waited for him for an hour and then two hours. The third hour we got a telephone call. It was Mr. Forrestal calling from Philadelphia. He'd forgotten that he'd been flown up there and had taken a train on the way back to Washington.

*Why do you think somebody would do that? Too many things on his mind?*

I'm pretty sure he was that way. He later committed suicide.

*Do you think he had Alzheimer's?*

No. He just had so many things going on in his mind that finally he went over the hill. He was in the Navy hospital in Washington, D.C. after the war ended.

*Give me another flight that maybe meant a lot to you.*

My first command flight was flying a Lockheed Loadstar, R-50, which was the Navy designation. I had taken several admirals, several captains and a couple of Army colonels and generals, and gone up to Portsmouth, Maine, to the Navy station up there. The story was that there was German submarines there that had been captured at sea. This was right at the end of the war in Europe. I later found out that this sub had uranium on board and was supposedly en route to Japan. Anyway, I took these men up who inspected it. This situation occurred prior to any public knowledge of nuclear bombs. We learned that the men we transported up were in on the nuclear fission program in Kentucky and Tennessee. I took them up there and brought them back.

The flight marked the only real accident I ever had in an airplane. I was landing and the plane was rolling too fast. I tried to make it turn and happened to hit a barrel on the runway.

*What happened exactly?*

We were landing at Anacostia. The repair crew was doing some runway work. I got cut off on my first approach by an SNJ, then had to make a go-around. I got going too fast on my go-around and, when I landed, I hadn't slowed down quite enough. I made a little bit of an uncontrolled turn and hit this barrel and ripped the bottom of the fuselage a little bit.

*Did that tick you off?*

Oh, it was very disturbing to me because I'd never done anything like that and wasn't supposed to do things like that. I didn't catch any hell on it or have it on my record.

*Were you a perfectionist?*

I tried to do the best I could. I didn't continue with my flying after the war because I didn't feel like I was going to get enough time to keep up my skills. I don't think you should fly if you don't keep your skills up. I checked out the Naval Reserve right after the war, but their equipment wasn't in the greatest of shape, so I wasn't going to fly that, either.

*Describe the kind of scuttlebutt when these guys were coming back from finding this uranium.*

We were forward in the cockpit, of course, and could not hear what was going on or any conversation. It wasn't until about ten years later when I was talking to a retired Army officer and I told him about that. He said he was familiar with the circumstances and what they were up there inspecting.

*So you didn't realize the implications of that at the time?*

Not at all.

*Were you privy to Winston Churchill coming in to Anacostia?*

Yes. He came in a Sunderland flying boat to a meeting with President Roosevelt at the White House. No information was to go out as to when he departed. In fact, they closed down the air station. Anybody who came on the station, whether a baker bringing bread for mess or someone delivering gasoline, were retained on the base for eight hours after Churchill's plane had left.

*Did you see Churchill?*

No.

*You must have done a lot of flying.*

Yes. We had 100-hour months on the duty I was doing. You take commercial pilots and that's something their contracts do not permit them to do.

*Talk about the conditions of that. Was that kind of exhausting?*

Yeah, at times it was. It was not unusual to make a round trip to California in the course of a week and then to do it again the following week. We weren't flying two-hundred knots. Instead, we were flying Navy R4D, DC-3s, Loadstars and Navy R-50, which had a cruise speed of around one hundred seventy knots. The R4D was around one hundred fifty.

*So how long would it take you to come across country?*

We'd usually stop off in Oklahoma. We'd take off at about 8 o'clock in the morning and be in Oklahoma around 5:30 or 6 in the afternoon, spend the night and then go from there. Sometimes, going west, we'd go past Chicago, and then spend a night in eastern Wyoming, and then cross the mountains. At that time, we didn't have pressurized cabins. If we got up to twelve thousand feet, we'd definitely use oxygen. At night, we'd always use oxygen. We didn't fly any more night flying than we had to.

*You were flying using beacons?*

Yes, we used the regular quadrants that they had on the airwaves that the commercial pilots used also. We also had automatic direction finders to tune into the station ahead, or the station behind, or to the side. The automatic direction finder was a radio you could tune to AM stations and receive an audio reply. The reply was converted to a compass dial with a directional needle. The needle would then point to the station you were tuned in to.

*Did you have any funny stopovers or any unusual things happen?*

I remember one time we were on our way from Washington, D.C. headed for Seattle, Whidby Island. We had just gotten over northern Indiana when we lost an engine and had to make an emergency landing at an Air Force base just outside of Fort Wayne, Indiana. What it turned out to be was a connecting rod, which had broken loose from a piston. Of course, that meant that the engine had to be shut down. That was a night landing. It was the first time that I'd been in an emergency of that magnitude. I was flying co-pilot at the time. I was alert enough to sense it. The pilot and I were very close enough in our thinking that we both hit to the feather the prop button at the same time. The next day the authorities brought us an identical plane from our base, and we continued our flight. Later, after mechanics had put the new engine in at Fort Wayne and the pilots took off with it, they lost the engine we'd been living on.

*Did you every stay at any farms or run into any farmer's daughters?*

When we were going into Los Angeles, we stayed in hotels and there were plenty of girls around who would like to pick you up. I never got involved. I was newly married at the time.

*You flew test pilots from time to time?*

That occurred on the trip we made across country when we lost the engine. We had these test pilots for the PV-2 (a Lockheed Hudson with two two-thousand HP engines) that were going out to indoctrinate the pilots who were going through training with them. We were flying along and then, all of a sudden, we were nose up, starting to gain altitude. At the time, we were flying autopilot so we turned it off and got it back into stable condition. Then a little while later, we were nose down. We realized something was happening. We had the door to the passenger compartment open. We looked back and saw about five or six of the test pilots were walking up and down the aisles. We reset our autopilot and were all right.

*Were they fooling around with you?*

They were.

*Describe that a little more.*

There were about five of them going together. Just one wouldn't have made any difference in our flight pattern, but when five of them, of say roughly one hundred eighty pounds apiece, did that it changed the trim. Depending on whether they were going fore or aft you nose would either go down or it would go up. You'd notice you were starting to climb or descend. So what we did is we snuck a look back and caught them the third time they did it. Then we pushed the nose over and then pulled it up quickly and bounced them off the ceiling. But it took us two ventures like that before we figured out what was going on.

One of my favorite stories involves Admiral Byrd, who used to come on board to get his four hours a month of flight time in. I think I'd just made Lieutenant J.G. at the time. I was assigned to take him up in a little twin engine Cessna Navy JRC, so he'd go and get his four hours in. Before we started I said to him, "Do you have someplace in mind to go?" He said, "Let's fly up the Shenandoah Valley." That we did. "Now," he said, "when we get past this ridge here, turn to the port and start descending." So I did. We got down and I said "How far?" By now we were down to about 500 feet. He said, "You see that farm out there? Get to 500 feet and hold it level. There's a nice house down there. Go over that and change pitch." I changed the pitch of the prop as we went over. He said, "That's where my mother lives." Then he said, "Now go back and land." We landed and a Packard with a chauffeur pulled up on the field where we landed. The field was not a prepared landing strip, but it was a reasonable field. He said, "This is my mother's chauffeur. We're going to go have tea with her." As a consequence, I met Mrs. Byrd.

*Describe the tea.*

We had some tea and cookies while sitting on the veranda at this nice old house there. At one point, Mrs. Byrd turned to me and said, "You know my sons, don't you?" I said that I'd just met the admiral for the first time but I knew she had two others. She said, "Yes, one is a senator and another is a governor." She wanted to know if I knew their names. I said I knew that Admiral's name was Richard. She said, "They're Tom, Dick and Harry." She was a very delightful older person, but she was probably not as old as I am now. She was probably in her seventies. As I recall, the Admiral at that time was in his young fifties.

We had gotten a new captain on board the air station. I was given the assignment of making the oyster run over to coastal Maryland. At that time, we had a little twin engine JRS. I had a co-pilot fresh out of the fleet who'd been a fighter pilot. He'd not been into any kind of twin-engine plane of any sort. Anyway, the captain wanted oysters and farmer's butter. It was in the wintertime. We went over to pick up the food, but on the way back we ran into one hell of an ice storm. The Cessna had no de-icing equipment and a very little heater in it. About halfway back, we began to ice up very badly. The windshield was iced over, and I was getting a little apprehensive about going in. At this point, my co-pilot had a cigarette lighter and was thawing a little hole in the windshield for me so I could see. Then I called in to Army Air Force at Bolling Field in Washington, which was right adjacent to the Naval air station. We had a connecting taxi strip with it of about five hundred feet. One had a three hundred fifty degree heading and the other had a three hundred sixty degree heading. I think Bolling had the three sixty heading, and ours had a 350 heading. With that connecting taxi strip, we had about 12,000 feet of usable runway, if necessary. I made an approach from the south but I was so iced up I didn't want to slow down. I tried to put the wheels down and the electric motor shorted out. We cranked the wheels down by hand, because of the ice. Then the flaps wouldn't come down. Finally, we landed with no flaps and a load of ice. I think I hit the ground at about one hundred ten miles per hour. That was one of the scariest things that happened to me while I was flying.

*So you made it in. What happened?*

I made it in and used almost all six thousand feet of the Air Force runway, plus some of the taxiway before I slowed down to a reasonably controlled speed on the runway. I had taken a lot of time because I didn't want to get down too slow. That plane stalled out at probably around forty-five or fifty knots, so I didn't want to get anywhere near that speed. I made my approach at about one hundred ten knots. When I touched down, I had a good half-inch or more of ice on the wings and fuselage. That increased the weight of course. The fact that I couldn't use the flaps made it touchier.

*So you were pretty scared.*

I was thinking very serious thoughts.

*Like what?*

What's going to happen? Can I keep it under control? I hope that I don't stall out in the turn as I come in for the approach, because I was coming over a residential area outside of the city. In fact, I lived not too far from the place.

*Did you lose your oysters?*

(Laughs) No. I got the captain his oysters and his butter.

*What did he say?*

I never talked to him. I just told my squadron commander and what I'd done. He looked at me and said, "Wanna do it again tomorrow?"

*He wasn't serious?*

No, he was just teasing me a little bit to see if I still wanted to fly. I said, "Well, I don't want to do that again, but I'll fly tomorrow."

*You flew down to New Orleans. Is there a story related to that? The Higgins Boat Company?*

The Maritime Commission was making a visit to the Higgins plant down there. They were making PT boats. Admiral Vickory was on board my plane. Admiral Land and Admiral Vickory headed the Maritime Commission. Anyway, we went in to New Orleans and went out to dinner one night. This was the second not the first time. The second time we went there we were with some congressmen, including the chair of the congressional house, Congressman Ebert. Afterwards, I was walking with him on the street in the French Quarter. I said, "Congressman, there are some guys following us." He said, "Never mind, those are just my bodyguards."

*What was the most unusual request a passenger made of you?*

I remember Admiral John McCain, who was the grandfather of the past presidential candidate, John McCain. He always had a Thermos and we surmised it must have been his martinis, because everybody knew he loved them. He was a very reasonable man to fly for. He wasn't anybody who pushed you hard. Everybody called him "Popeye" because of a squint and bad false teeth. Another time we had gone up to Long Island. With us, we had Artemis Gates, who was the Undersecretary of the Navy at that time. His home was there. We dropped him off on a Friday and went back on a Monday morning to pick him up. In the course of waiting and waiting, he finally got in and came up. After we took off, he came forward and wanted to apologize for being late. He said he ran into a problem getting on board the air base there. He explained that he was in his own car and his chauffeur was driving him, and the sentry met them and challenged them. He said he was Artemis Gates, Undersecretary of the Navy and was to fly out of there, and the sentry said, "Sorry boss, this is the Army. You don't belong here." So he'd been delayed. He was another gentleman. I was impressed with him, Mr. Gates and Mr. Forrestal. They were all very considerate of the pilots. In fact, our unit, I later found out, was given a citation from the Secretary of the Navy.

One amusing incident occurred on Christmas Day, 1943. I had stand-by duty in our flight operations squadrons along with three other pilot officers. The routine was that we were to be on the base at 0800 hours in the morning. I showed up on time; however, there was emergency activity going on all over the base. I went up to the pilots' ready room and it was filled with all the pilots in our group, all of whom were in flight gear. Those of us on duty had not been called—just the off-duty pilots. Someone told us that naval intelligence had been informed that a German aircraft carrier had been sighted off of Cape Hatteras, and that all Navy and Army Air Force bases were on red alert.

At Anacostia, we had several F4F, F6F, and F4U Navy fighters, which were all armed and prepared to go. Along with these fighters, we had about twenty SNJ fighter trainers. The SNJs would have a single 30-caliber machine gun fixed to fire forward, and a flexible gun in the rear seat. There were prepared to help defend Washington D.C. I was assigned to one of the Js.

We sat around in the ready room until about noon when we were told all was secure, so we didn't need to fly out looking for a carrier. It turned out that an officer in naval intelligence had done too much celebrating on Christmas Eve and thought he would test the East Coast defense preparedness. I heard later on that he had actually been court martialed.

Most of the pilots in our group were fleet returnees. One had been at Pearl Harbor on December 7; two had flown PBYs from San Diego to Pearl; and one was on Johnson Island, just south of Hawaii. Another was in the Philippines that fateful day. The pilot at Pearl received a Navy Cross. He had taken off in the middle of the attack and attacked the Japs. The one in the Philippines left there shortly after the attack in a SOC (a single-engine floatplane) flying only at night. It took him several weeks to get all the way to Australia. He ended his naval career as a Vice Admiral in command of the Pacific Fleet.

There were other pilots in our group who also received decorations. Among them was a Congressional Medal, another Navy Cross, a Distinguished Service Medal, Several Distinguished Flying Crosses, Air Medals, Purple Hearts and, most unusual, a British Distinguished Flying Cross.

*What was the unit called?*

It was Flight Operations, U.S. Naval Air Station. We were just the operations squadron out of Naval Air Station, Anacostia. We had two fatalities while we were there. One had been an old time Navy chief, who was commissioned as a lieutenant and he was giving an instrument flight check to the squadron commander. They had an engine problem and crashed and the chief was killed along with an enlisted man who was the mechanic on board the plane. Our commanding officer, Commander Gordon Fowler, USN, was also very badly injured. This was in the last three months that I was on the base. I was back in Washington four or five years later and looked our commanding officer up in the phone book and gave him a call and he was working as a stockbroker at that time. His father had been a brigadier general in the army and this man had gone to Annapolis.

*Was anybody jealous of you? Did anybody ever infer that your choice was a "plum" and that you guys had it made?*

I never heard anything while I was there but since have known that there were people who wondered how I got the cushy job.

*But you didn't want the cushy job.*

No, when it came to transfer, I put my service in there and happened to be with a commander who had my portfolio on his desk. He was in Navy personnel. He said, "You're coming up for reassignment. Do you have anything that you want?" I said, "I don't think I want to go into Naval Air Transport. I've had that." He said, "You have your choice of Naval Air Transport or night fighters off of a carrier or going to a PB4-Y patrol plane commander's school." I took the patrol plane. I didn't want any part of night fighters. I'd spent one day on a carrier while they were doing practice landings. It was an escort carrier in the Chesapeake Bay, and we were out there with a bunch of people who'd gotten a Navy "E" for their efficiency. I would say that ten approaches were made to that carrier and only one actually went down on the deck. The others were waved off, but one man ran into the stern and hit the control tower.

*But you wanted to go to the war nonetheless.*

I had another admiral I knew named Admiral Miller. I got acquainted with him and got an assignment to take him out two or three times. I've got a feeling that he requested me after the first time because on the third time we were out, he was in the cockpit and he said he was to be reassigned to London as the Naval attaché. He said he needed an aide and a pilot. He said, "How would you like to come with me?" I said, "Admiral, I haven't been out in the war yet and I think I should do that." He said, "I think that's the right answer. I'd be willing to take you with me, but I do think you're giving the right answer."

*That didn't take soul searching?*

No. I had two years of cushy service and an awful lot of friends who never had even the smell of it.

*Did you lose any friends?*

Several. Right out of flight school, there were guys lost. Even in flight school there was one, not a real close friend, but an acquaintance. After one of my good friends, who was at my wedding, got his commission, he went up and ended up at Guadalcanal and we lost him there.

*Do you remember getting word of the loss? Did it make you really want to get in the war?*

It made you think. I can't say it made me want to do more than I had been doing. I knew that my time should come to be in that circumstance.

*How do you look back on it now? Obviously, you've contributed a great deal. Is there always a lingering wish that you could have been in battle?*

I just consider that I had a tremendous break. I also feel that I got the break because I had the skill to take it. In academic standing, I was in the upper half. I know I was in the upper half going through. I know I was one of the few guys who went through with only one down through the whole flight training program and I knew that I had that skill.

*Biggest compliment you ever received?*

I think it was from this admiral who wanted me to go as his pilot and aide. He recognized that I had the ability to do that. Then at the end of the war, I interviewed with United Airlines and they offered me a job because of my background. I had a lot of flight time, nineteen hundred plus hours, and an instrument rating and had flown airways in the United States, which is what they were looking for. I chose a job in engineering at General Motors, and they paid me almost as much as the commercial airline was offering me as a copilot. That would have been a cut in pay for flying. I would have been away from home just like I was when I was in the service. My wife and I decided that wasn't good for being married. We had one child at the time and another on the way. That ended my flying career.

# Fighter Pilot and the Checkerboard Clan

## Chet Sluder

Chet Sluder loved flying so much that he quit eating, at times, while a student at the University of Texas. His goal was to save the six dollars a week, so that he could buy 30 minutes of flying time. Chet was taught to fly at Austin, Texas, in a Waco 10, OX5-powered, by a local pilot, Don McDonell. He spent the summers between school years at Stinson Field in San Antonio, at the hanger of Dick Hair, who had been taught to fly in 1916 by Margorie Stinson, on a Wright Model B. Sluder would solo on January 12, 1933 after little over seven hours flight time. He flew his solo, he claims, in a "total trance."

A fighter pilot was what Sluder turned out to be, and he flew most of the fighter planes, as he explains in the interview, from the early thirties until the sixties. During his illustrious career, he headed three squadrons of fighter pilots who escorted bombers during WWII.

Sluder entered Randolph Field in June 1935, then graduated from Kelly Field in June of 1936. He joined the Ninth Bomb Group at Mitchel, then went to the 8th Pursuit Group (not Fighter, but Pursuit) at Langley in 1937. He returned to Kelly to instruct in his favorite P-12s, then went to join the 12th AF in North Africa. He worked in the same building with General Jimmy Dolittle and recalls meeting him passingly. He says, "I worked inside his operations section in Algiers at headquarters in a large oil company building. It had a huge staircase and anytime I saw the general he was going up and down two steps at a time."

Sluder joined the 325th Fighter Group in July of 1943 on temporary duty, then joined them permanently in September. In December of 1943, he moved to Italy and started flying missions from there. In April 1944, he took over command of the Checkertail Clan, whose planes' tail surfaces shown with a yellow-and-black checkerboard.

Sluder retired in 1965 from the military and then flew for Carco and Ross Aviation for nine years. His job involved hauling passengers and freight for the Atomic Energy Commission. Finally, he retired from flying altogether after some 13,000 hours, in 1978. He now lives in Albuquerque, New Mexico, with his wife Zimmy and involves himself with golf and computers.

Now in his eighties, Col. Sluder is a small man, with twinkling eyes and a clear-cut wit. Despite his size (5'5"), he is a nonetheless a commanding presence and a strong and proud voice of the fighter pilot, as becomes energetically clear in the interview.

*Any stories about dealing with your students in the early days?*

Yes. In 1939, there was a program to pump civilian pilots through flying school in three months. Joe Mackey, who was a civilian pilot before WWII, was one of those students. At that time, I didn't know who he was. He was just a student to me. One day, the section chief told me to get a couple of P-12s and take Lt. Mackey up, and check him out on aerobatics. So we started out toward the aerobatics area, climbing up, and I looked back to see how this student was doing in formation. He was holding good position on the left wing, but inverted upside down. (Laughs) We got up to altitude. I gave him the signal to get behind me. He did and I did what I thought was a pretty good, slow roll—an aileron roll. Well, Joe followed what I did by doing an eight-point roll. Then I did a loop, and I thought, "That was pretty good." So Joe did a loop, except he stopped up on top, and did an 180 degree inverted turn and came back down the way he'd gone up. So I backed off and let Joe put on an airshow. (Laughs) See, I didn't know until later that he had had his own airshow prior to school, and that he had flown with Roscoe Turner in the National Air Races. At that time, the CAA required you to have your license renewed annually, so one time Joe and I took a couple of P-12s from Kelly and flew over to Stinson Field in San Antonio. The CAA man was going to be there. So we got our tickets renewed and came back to Kelly. Coming back, Joe was on the wing, flying in good tight formation. I gave the landing signal, then we did a nose-high 90-degree turning slip. When we got down on the runway, Joe's lower wing tip was about four feet from my stabilizer. He never moved from that. He hit four wheels, and the two tailskids hit the ground simultaneously. It was a nice landing. But the base commander saw it and wanted to know who those pilots were. (Laughs)

*You've pretty much spent your life in fighter planes. Do you have a favorite?*

For just plain fun, I have to say the P-12. It was a little light airplane, had a pretty big engine in it, so it had real good performance. An open cockpit, bi-plane and fixed gear. No flaps. It was very responsive to the controls and you could do lovely aerobatics with it.

As for unusual fighters, maybe the P-35 was a little bit unusual. The thing I remember about it was that at high speeds, the canopy shook like crazy. All the planes I flew seemed to be quite gentlemanly airplanes.

*How many planes have you flown overall?*

I've flown about 130, if you count the modifications over the A, B, C, D, etc. However, I've been fully checked out on 87 different planes. Among the fighters I've flown include P-1, P-6, P-12, P-35, P-36, P-38, P-40, P-47, P-51, F102, f-80. F-84, Spitfire, Hurricane. I've caught rides in the F101 and F106, but I didn't check out in them.

*Did you ever have any forced landings?*

Never did. I lost three engines flying DC-3s at three different times, twice on take-off, and once while cruising. I guess that's a forced landing, isn't it? But I had one engine left.

*Where'd the name Checkertail Clan come from?*

The group commander, Bob Baseler, was in North Africa before I joined the outfit. I guess he had a lot to say about a WWI Ace from Germany, named Voss, who had a checkerboard painted around his fuselage, on his Fokker. Baseler liked that checkerboard idea. Baseler came up with the idea. He tried black and white and red, but finally settled on a black and yellow, which showed up better than the other color combinations.

That became our trademark and, surprisingly, since the end of the war, even to this day, I can find bomber crews who thought the checkerboard was the 332nd, the all-black group, the Tuskegee fliers.

*Why do you think that was?*

I have a theory and I think it's probably valid. We gave the Tuskegee fliers our P-47s very quickly. That is, they were transferred very rapidly, when we were getting our P-51s. So, we didn't have time to paint out our tails. I think some of the Tuskegee fliers got to landing that P-47 on checkout, around Italy, and there was that checkerboard tail.

*Did you know any Tuskegee fliers?*

Yes, Col. B.O. Davis. I did know some stories. There's one about a guy who gets lost on his mission, so he calls up on his radio and says, "Come on down. I've got two or three Me-109s down here." The response he got, according to the story, was, "You got the Distinguished Flying

Cross, why don't you do some distinguished flying?" I have no idea whether it's true or no. In fact, I doubt it.

*What was you most successful attack?*

On January 30, 1944, we got every airplane we could up in the air and flew on the deck with sixty airplanes up the Adriatic Sea to northern Italy, then we climbed up and out over the German fighter fields, just as they were scrambling to intercept the bombers that were about 20 minutes behind us. We got 37 enemy aircraft that day. Our top guy, Hershel Green, now AF retired, got six himself.

We were on the deck and the word was, if you lost an engine, don't bother trying to get up high enough to bail out, because you'd get up in German radar. You'll just have to go into the drink. We were so low we were literally leaving wakes in the water from our prop wash. We might have been ten feet off the water.

Sitting up there, Green saw a bunch of JU52s, and so he attacked and got four of them; he also got a Macchi AMC 202, an Italian fighter, and a Dorner Do217, a medium bomber.

*How many guys did you lose?*

We lost two. The two guys that we lost ran out of aerial targets, so they strafed a German airfield. They thought it was so much fun that they turned around and did it again. On the second pass, the Germans were waiting for them and knocked them down. Both survived, though one of them, Fallon was later killed at Eglin AFB. The other was Lowery, whom I've lost track of.

*Any other missions that stand out?*

About the time we got our first P51s, we led escort for the first shuttle-bombing mission into Russia. We took off from Italy in four B-17 Groups, then flew across Europe, bombing a target en route and then landing in USSR. It worked like this: We picked up the bombers for escort over Yugoslavia, and carried them as far as the Dnieper River. I left them there and headed for my assigned base, Piryatin, in the Ukraine.

That country is flat as a pancake, It was real good black farmland with no landmarks. No rivers, towns, mountains, nothing. I was doing time and distance calculations off of three maps – three different scales, three different color schemes, put out by different people. I finally figured, "By golly we've had long enough to get there. We must have overshot it." So I did a 180 degree and headed back to where we'd been I was planning to go back to the Dnieper River, and then follow it to Kiev. I figured it wouldn't be too far, and that we could get on the ground there before we all ran out of gas. While I was going back, I looked down and saw a P-39 and a 2-1/2 ton, GI truck. I didn't know where we were, but I knew that was where we were going to land. It turned out that's where we were supposed to be.

*Is that in your book, "Secrets of Navigation?"*

Yeah, it's spelled "L-U-C-K." (Laughs)

*Tell me more about the bombing.*

Our target en route was Debrecen, Hungary. One B-17 caught on fire up there, but everyone got out of it. No enemy action there, though. We went in and landed. While in Russia, we flew one mission into Romania and back to Russia. On that mission, I remember seeing two planes climb up toward the bombers, one behind the other. I dropped my auxiliary tanks and went down, and deliberately overran the wingman, to get to the flight leader. I was thinking my second flight would get the wingman. I was coming in from an angle. I checked my aim, squirted again, missed, squirted again, and hit right behind the cockpit. That's the first he knew I was there. When he saw me, he started to brake, and I squirted again and got him right through the canopy. He went on to the ground vertically from about ten to twelve thousand feet, but he got out of it all right.

We got six Fock-Wulf 190's that day. Everybody got back and landed in Russia. The bombers were really happy because it turned out to be one big, milk run for them. They'd gone and hit the target and didn't get any flak and didn't see any fighters. I had to tell them, "We saw the Germans first. That's why you didn't see them."

Incidentally, there's a newsletter put out by one of the bomb groups, where one of the pilots went to great length to ridicule the fighters. He said they were worthless; they didn't do any good; they never were where they were needed, and that all they wanted to do was become aces. Yak. Yak. Yak. I felt called upon to write a response.

*How many planes did you get?*

A couple. I also got a Me-109. We were going up Italy, escorting B-17s and ran into a bunch of Me-109s. I saw some make a head-on pass at the bombers and then start down, and I turned in behind them. I caught up with them. My doggone wingman found something else to shoot at, so he didn't stay with me. So I caught up with these guys, and I wound up on one of the wingmen. I squeezed off some shots and immediately was on instruments, because I hit his coolant, and I was in the middle of a cloud of his coolant. I vaguely saw him in the cloud, so I squeezed again, and hit him some more. Then I had to peel off, because this guy's flight leader was beginning to get irritated. He was turning back, trying to get on my tail. So I had to stop and work with him for a while.

Neither of us did any good. I was in a P-47 and as we fought, we came down. By then we were at about 10,000 feet. Anyway, the old gal I flew didn't turn so well down low like that. So he would stall and spin out and I would stall and spin out and recover. We did that for a while longer. Finally, he pitched out and headed down, and I let him go.

*Why did the P-47 have problems?*

I don't know, I'm not an aeronautical engineer. But, up high, the P-47 was a good fighter, with a fast rate of roll. Down low, it was heavy. It just didn't turn as well as some airplanes did. Another thing: at high altitude if you rolled over and started down, you got into compressibility. Your controls would lock up. There was nothing you could do except sit there and hold pressure back on the stick with full throttle, until you got down to low altitude, with high density air. At low altitude your controls would become effective and you could pull out of a dive.

*Did that ever happen to you?*

Yes, one time. I was chasing a target at full throttle. He was trying to head home. The next thing I knew I was just sitting there, locked in. You couldn't do anything about it.

*Was that the most frightened you've been? You knew about the problem, but it's got to be scary, nonetheless.*

Well, it's a funny thing. I don't know if I'm typical or not, but the word "frightened" somehow doesn't seem to fit. It should. I mean, anyone with a grain of sense is going to get frightened. You just look at the situation and do what you have to do. Honestly, I don't remember ever being scared. I was attentive on certain occasions, depending on what the circumstances were.

*Any particular lesson you learned in the war?*

Keep lookin'. (Laughs) As a matter of fact, I'm not very big, as you know. But my neck size increased a full inch during the war. I lost weight and was down to 112 pounds, but my neck size increased. You kept you head on a 360-degree swivel whenever you were on a mission.

After the war ended, I had just come back home, and was stationed at the Pentagon. One night my wife and I were coming back from having dinner out, in Washington. I was driving along, and I looked in the mirror for any Me-109s. Without thinking, I said, "You know, the last two years have been the happiest years of my life." Boy, did things get colder in the homestead, after I said that.

*Any special compliment you've received?*

When a mission came in on the teletype at night, for the next day, the operations officer would come by my tent and tell me what the mission was and ask if I wanted to go or not. Anyway, several years ago at my reunion I was really flattered and embarrassed, when my old operations officer told me the thing he remembered most about me had to do with the missions I went on. He said, if the mission involved going across the Adriatic and didn't amount to anything, I'd tell him I didn't want to go. But, he said, if it was going to be to Vienna or Ploesti, I'd go. He said I'd go every time if I thought it was going to be a hot mission. I thought that it was nice of him to say that, but as I said, it embarrassed me.

*So you flew to the oil fields in Ploesti?*

Yes. It was just a routine mission, except that there were lots of defenses over there, so it was a good one to go on. It was smoky. The Germans used smoke to cover the targets. We were escorting the B-17s and B-24s over there and got in some pretty good scraps. I missed one of the better ones over there, though. The pilots ran into a whole lot of Germans and claimed 18 enemy airplanes.

When the mission came back, the phone in my tent jangled. It was one of the bomb group commanders. He was mad and he said, "It's a good thing we didn't see any German fighters because your men left us undefended." (Laughs) What happened was our pilots saw these vapor trails and found all these Germans. We went over and proceeded to mess them up. But this commander didn't know what happened. He was real unhappy. I told him to go to hell.

When we were in Italy with the P47s, we'd get these teletypes from headquarters, and it was clear they didn't get it. The teletypes would say, "Pick up the bombers at this point and carry them to this point and bring them all the way home." We didn't have that kind of gasoline. So, we'd argue with them. We'd say it's different cruising between Kelly Field and Wright Field. We finally quit arguing and did our escort as we saw fit, doing the best job we could.

Incidentally, because we went a lot faster in the fighters than the bombers did, we would "S" across the bomber's stream, going perhaps five miles to one side and five miles to the other, just back and forth. One day we were "S" ing" back and forth routinely on the way to Budapest. All of a sudden, on the radio someone said, "We're being attacked. We're being attacked. Where's our escort?" I looked around. Gosh, there wasn't anybody else there, except us and the bombers. The weather and visibility were perfect. I looked some more, but there wasn't a German in sight anywhere. Then I noticed every time I would "S" over the bombers, the top turret would shoot at us.

One day as we headed into Budapest in P51s, I saw six vapor trails coming real high. Way up there. I called up my people and did a fast 180 to head the same way they were going. I was a little late getting down. They hit the tail end of the bombers and knocked down three B-24s in one pass. Five of these six pilots split-S'd, hit the deck, and went home. The sixth guy, I guessed it was the flight leader, pulled up parallel to the bombers, but out of their range. Then I saw him coming toward me. I tell you what, I thought, "This is a cinch. I almost feel sorry for the guy." So I turned in behind the guy, but I didn't get underneath his trail. And just as I squeezed, he saw me. Boy, he did the damnedest things with that 109—things that no airplane could possibly do, but he did them. I never touched him. But he never touched me either.

Several years ago, we had a big party out at Las Vegas. I went to it and was walking around. I sat down by one of the exhibits with four or five guys, to rest a little, and got to talking. One of them introduced himself as a Hungarian Me109 pilot. So I said, "I flew P-51s around your hometown there." And I described that particular incident to him. All of a sudden, he said, "I was in that 109." He is convinced of it, and I can accept it. He had the time and date and circumstances pretty doggone down pat, so he may be right. His name is Michael Karatsonyi, and since then we've become good friends and he's flown over to visit three times.

*Remember your last flight?*

Yes, we were going up to Markersdorf, escorting some B-17s. Before we got there, we had a whole bunch of Me 109s and Focke-Wulf 190s come in on an attack. One of the 190s peeled off and started down, and I started down behind him. I quickly called up my wingman and didn't get an answer, so I looked back to see where he was. When I did, I looked right into the nose of another Me109. It had already started firing. I could see the tracers. So I contemplated that for a mini-second, then started to take some action. He didn't hit me. This guy could fly well, but he couldn't shoot well. In fact, he was a lousy shot.

A P-51, if it gets going fast, will start to roller coaster. You go up and down like this. If you go a little bit faster, it will start to roll. I just helped it roll. I had everything against the firewall. Then I straightened out, and there would be a string of bursts from the same 20mm. Once on the way down, I stopped a roll and I was upside down. The thought flashed through my mind, "I don't have any armor plate on the bottom of my seat." And this struck me as funny (Laughs) I started chuckling right in the middle of a dogfight, and this guy was trying to kill me!

I got down to about 10,000 feet. I did a 180-degree turn and came back out and, as I did, I saw this same guy come by me with four of my men on his tail. They got him. That was my last mission.

# Sinking the Yamato

## Chandler Swanson

General Swanson Wartime Fleet Pilots 7-26-43
Chandler in Back Row – Second From the Right

On April 7, 1945 in the East China Sea, Chandler Swanson led a group of torpedo bombers on one of history's most successful torpedo bomber missions, sinking the Japanese ship the *Yamato* which was headed on a suicide mission to Okinawa. The *Yamato*, a 72,000-ton battleship, was the largest battleship ever built. The U.S. Navy referred to it as the "biggest battle wagon remaining afloat under the flag of the rising sun." Swanson got the first torpedo hit on the *Yamato*. Overall, during the five-month campaign in the South Pacific, Air Group Eighty-four, of which Swanson was a part, destroyed 234 enemy aircraft and 160,000 tons of enemy shipping. Swanson also damaged major plane factories in Tokyo. After the war, President Truman wrote to him, "We now look to you for leadership and example in further exalting our country in peace." Among others, Swanson received the Silver Star with gold star; Distinguished Flying

Cross with gold star; the Air Medal with three gold stars; and the Navy Commendation Medal.

Swanson was born in Brooklyn, New York, and grew up in Montclair, New Jersey. He joined the navy in Minneapolis, then went to Pensacola in 1937 and 1938. He was a member of VS-2 from 1938-1942: NATC from 1942-1944. Then he joined VT-84 for just over a year and VFB-156 at the end of the war. Swanson returned as staff at Quonset point, and served on the *USS Kearsarge* for one year. After a variety of duties, including Naval War College and Naval mission to Brazil, Swanson returned to the Naval War College and later commanded the Naval Air Station in Sanford, Florida. He retired as captain in 1969, having maintained his flight status throughout his thirty-two years in the Navy.

Leaning on his cane, this modest eighty-five-year-old gentlemen greets me at the door of his Florida high-rise, then escorts me in. He shows me his cabinet of his prize minerals and introduces me to his wife, Ruth, whom I learn was at Pearl Harbor having breakfast when the Japanese struck there. He offers me a beer, and the interview begins.

*How did you get into aviation?*

I was thinking about aviation all my life and I figured the military was the best way to do it. In 1927, when I was about twelve, I followed Lindbergh's famous flight. I admired the way he navigated. Later, I attended Carleton College in Northfield, Minnesota. One day after I graduated, I was in the Navy.

*Was that a shock?*

No, it wasn't.

*What was the key to your attack on the* Yamato*?*

The pilots in the squadron were the key. We had a lot of things. A cooperative ship—the *Yamato,* for one thing. The weather was really bad—with a ceiling of something like nine-tenths overcast. I found a hole in the cloud cover and we attacked and dropped our torpedoes. Flying was treacherous. AA was beyond belief. We lost one plane, which exploded.

*Do you remember whose that was?*

Yes, Walsh was his name. Guttenberger's plane got hit and the gunner was knocked cold. I think Webster's plane was also hit and was burning, but the fire went out. We were crisscrossing the target, which was tricky. Some of the planes could have easily collided. You know that we weren't the only ones there. The Marines had a Corsair squadron there, too.

*So then, your squadron dropped your torpedoes.*

Yes, most of us hit on his port side. For the attack, the squadron split almost fifty-fifty. I had weight in one squadron. There was no voice communications necessary, because we'd done it so much that we knew what we were doing. Part of the guys went left and we went right and came down. The Avenger torpedo bomber has a 2,000-pound explosive on its belly, so you're not going to play around. You're going to do what your business is.

A torpedo attack is best made head-on from the forward part of the ship. I said the *Yamato* was cooperative. It turned around unlike any other ship in the formation and headed north—whereas the other ships headed south. I didn't know that coming down through the clouds for two bow-on shots. We had nine hits out of thirteen attempts. Earlier, we had wanted to be on the far side, so we would be headed home, rather than having to come back through the whole gang of Japanese ships. And that worked just exactly right.

*Did you rendezvous?*

Once we got clear of the opposition—all the destroyers except one were sunk—there was no need to go fast, so I cruised slowly. We were flying under the overcast, which was about two-thousand feet and went back to the *Bunker Hill.* I didn't look back for one second for explosions. It took time before we would know what the details were. We were 250 miles away. It took about an hour for the Yamato to sink. We got the word back on the ship that it had gone down. When a ship gets eight torpedoes in the hull, it's a goner. We knew it was just a matter of time when we got back to the ship.

*Did you celebrate?*

To me, it was just a standard routine night.

*You must have been excited.*

No.

*But that's a major accomplishment—sinking a 72,000-ton ship.*

I was busy in the ready room getting dope on who got what hits.

*What was the closest call you had?*

I've had a couple of them. We were flying off the Bunker Hill and about a month after hitting the *Yamato*, we were hit by kamikazes. I was below deck when two of them hit mid-ship between the bow and the stern. Ensign Bacon was a name I remember. He got knocked down through a hole in the bomb bay and landed on the hangar deck. We lost hundreds. In fact, several of our squadron members were killed in the attack. I was earlier on the *Lexington*, in the battle of the Coral Sea, when it was sunk. I climbed off by shimmying down the line on the stern. I was in the water for awhile and was among the last to get picked up.

*Were you in a raft?*

No, I was in the water. I had my hands on a line that was hanging down.

*That had to have been scary.*

It wasn't scary. You're too busy looking for boats and signaling.

*How far away were you?*

We were right under the stern. I saw men drown there.

*They lost the rope.*

Yes, or got nervous and started to swim.

*Your drop on the* Yamato *was one of the most successful torpedo-drops in the war.*

It's amazing that we would find that many lucrative targets. There aren't many to compare. I'm surprised at the lack of recognition or write-up on this particular item. It's gone and by, but we got the *Yamato*.

*There were a lot of sailors on the Yamato.*

I've read somewhere over thirty-two hundred. I think three thousand or so were lost.

*How would you like to be remembered?*

I did what I was supposed to do and the planes did what they were supposed to do. And the *Yamato* did what she wasn't supposed to do. And we had a very good result.

SBC4s. *Lexington* in distance.

*What are you proudest of?*

My overall service was pretty unblemished. I can't think of anything I did that was absolutely wrong.

Aircraft Carriers Lexington and Ranger Taken From

the Enterprise about 1933

# My Early Flying in Idaho

## Bill McGinnis

Bill McGinnis's family had recently moved to Idaho from Spokane and was living on the homestead. In his new, quiet environs, he had lots of time to read, and read he did. The subject he found himself gravitating toward in the publications was flying, so at night, he would read about flying. That is where he claims he learned to fly. He adds, "Believe it or not."

*What magazines?*

Flying magazines. Of course, I was always nuts about getting ahold of an airplane. I had a friend in Spokane that I had met in 1922 in the Marshall Wilderness. He was twenty years old, and I was twenty-two. His name was Carl Shermer. I didn't know anything about what his flying or what he was doing with it. But I ran into him at Felts Field in Spokane. He had just bought an Eagle Rock, which was a new airplane. The engine was World War I surplus. He began to barnstorm. His father had died and he needed a few dollars.

You could buy stuff cheap back then, stuff for practically nothing. Twenty-five dollars for a prop that would cost you $1,500 or $2,000 today.

*So you did barnstorming in those days.*

That was a common thing back then. If you had an airplane and wanted to make a few dollars with it, you could. People liked it.

*What attracted you to flying?*

You had to apply yourself to get anything out of it. There were fellows who would take an airplane flight and say, "I don't want any more of that."

*How many years did you fly?*

I flew from about 1927 to World War II. I had a commercial license at that time. I had gone back and taken about 100 hours or so for it. We were thinking we would probably be called back into the Air Force because we had the time and the experience. But when it came to a showdown, I was 43 years old and they didn't want me.

*How did you feel? Did you feel sad?*

Well, yes, of course I did. I had five-and-one-half year's prior service in WWI, and military service in the Washington National Guard, and I knew eventually that I was going to have to go back into the service in WWII. I let them draft me and I went back in, of course. The military picked me up and sent me to OCS.

*Were you considered a daredevil when you were a young man?*

I would think probably I was considered so. As far as flying was concerned, I never got into an airplane—a strange one—that I didn't fool around with a little bit to find out how it flew. I still maintain that the old planes were the best planes in the world to learn to fly in. I mean, you could do things with them by way of handling the controls knowing that you were the boss of the airplane. A lot of fellows never learned that. You would find old-timers, fellows that were older, maybe with a little money, who would get upstairs and would get into trouble and not know what to do. And they would crack up.

*Tell the dog story we were talking about.*

The plane was sitting right out here in the district one time when we were barnstorming. The dog walks up and tries to bite the prop. He got his nose clipped. It cut his nose.

*Do you remember your solo flight?*

I remember it. I made three shots for the field before I hit it. When I landed on that solo flight, I made a couple of bounces and I jiggled the stick and rudder and tried to do everything to it to get it settled on the field. Finally, I got it down and took it back to the hangar.

*What was the strangest plane you ever flew?*

I don't know. I never really had any real trouble with any airplane, except for one. That was at Felts Field, too. A long time after I soloed, I was flying an Eagle Rock with a Hisso engine on it. They took the Hisso, a water-cooled engine, off it and put a J4 on it. The J4 had more horsepower, but was lighter.

They had trouble rigging it with that lighter engine. I took off with my brother from Felts Field one day and I hadn't been warned about it. We came in to land and the thing just wanted to float all the way across the field. It wouldn't settle down. And I told the owners about it and they said, "Oh, yeah, we know about it. We've just got to do some different rigging on it." But, before they got around to doing that, Carl Shermer and I took the airplane up with a big, heavy passenger. He owned part of the airplane, but had never soloed. This was for a spin inspection. Anyway, Carl took him up to give him some instructions. He finally got into a spin and Carl told him to get out. But Carl got no response from him. Carl said later that he thought that the fellow had both legs on the same side of the stick and couldn't move to control the plane.

Carl finally brought it out of the spin and, when he finally came out, he came out with his nose way up in the air. But he regained flat spin. That means without the nose being up. Once he crawled out on the wing, trying to bring that high wing down. The fellow didn't know what was happening. Finally, they came in and crashed. I've got a picture somewhere of that airplane. Finally, when it settled down, its wings came right down on the ground, flat.

*But they didn't get hurt badly, did they?*

The most serious thing that happened was that Carl lost an eye. His head went up against the firewall. It took him out of the flying business from then on.

*Do you remember anything about the flying magazines you read?*

These magazines were all technical matter. All of the flying magazines then were promoting flying instruction because everybody was wanting to learn how to fly. And, by learning the theory of aerodynamics and flight, what made an airplane stay up there that helped. I never knew much about navigation and meteorology. Those two things you need to know about, but not as well as communication.

*Any stories about Bob Johnson. He was kind of a legend, wasn't he?*

Bob Johnson pioneered smoke jumping in this part of the country. He and his friends, who owned planes, would get together and hop passengers to pay for the planes. One time Bob Johnson had this airplane on the field and Carl Shermer had the Hisso-Eagle Rock he owned out there. Shermer's was leaking oil. Carl didn't know about it, and Bob Johnson did, but he wouldn't say anything about it. He was waiting for Carl to find out, and quit hopping passengers.

*What happened with the leaky oil?*

Well, they discovered it before they took off, and took care of the leak.

*Sounds like a rivalry between the pilots.*

Well, it was.

*Problems flying?*

The closest thing to a problem I got into flying back in those days happened in Missoula. Back in those days, you could do as you pleased, kill yourself if you wanted to.

I liked to fly this J69 Travelaire. There were fliers around here who didn't have airplanes who wanted to get ahold of one to fool around with it. It was common to spin and loop and fool around to see how the thing handled, especially at slow speed. If you tried to stop in the air, you would stall. I took a friend, a flier and mechanic for National Airlines upstairs one time in Butte. So we got up and started to fool around and I said, "You want to fly upside down." He said, "Sure." So up we went and, when I put the stick in to keep the nose up in the air, the gas tank came off right in front of the front seat. I was flying from the back, though, and the gas—just a shower of it—came right down past us. Any little spark could have set us on fire. That's about the closest I ever came to a crack-up. Of course, when I saw the gas coming, I rolled the plane over. It could have put us in real trouble.

Before I soloed, Carl brought his airplane up and we were flying off this field behind Dike's River Station. We used to camp down there about 100 yards from the highway. Anyway, once Carl was flying and I was in the back. And I remember coming down over town. We weren't supposed to, of course, and we buzzed the garage. Right down over city hall and we buzzed it. I mean, we buzzed it! And I remember everybody in town talking about the racket.

One time we came in for a landing on this big long field. I thought we went too low. And the first thing I do is see these high wires in front of us. There isn't anything I can do, but I go right through them. As I did, I yanked back on the stick a little bit and the airplane just almost stalled there in the middle of them. That's about the closest I ever came to a crack-up with those three big power wires.

*Any sparks?*

I didn't see any, but the plane had a wooden prop, and he was a little worried about the little nicks in the prop— also in the outer face thrusts. So he called Spokane and, at the time, this Ford Tri-motor had just come into Spokane. Vern Buckwalter brought the Tri-motor up to bring that prop for us. That's the first time a Tri-motor was in Bonner's Ferry.

*Early Days, Felts Field, Spokane*

# Hitchhiking to the Heights

## Robert Goldsworthy

Robert Goldsworthy graduated from Washington State College (now WSU) in Speech Communication. He took a job at KFPY Radio in Spokane, Washington, as a radio announcer, but that didn't last long. "I wasn't very good," he says, "and I got fired." Feeling humiliated at losing his very first job, he wondered what he could do. His brother, at the time, was in the Army Air Corps. He noticed in the paper that the flying cadet board was at Spokane, at Gonzaga, taking applications. He hitchhiked to Spokane and walked out to Gonzaga where he took the exam. At the time, he didn't tell his folks he was accepted into the program. He was called up to Glendale, California, for adjoining primary flight training and then went to Randolph Field for his basic flight training. He ended up in San Antonio, for his advanced training where he was commissioned as a pilot in the Army Air Corps.

An early volunteer to fly the new B-29, he ended up in the Pacific Theater. He was on his third bombing mission when he got shot down, ending up as a POW for a period of nine months. He later flew bombing missions during the Korean War. A Command Pilot, he retired with the rank of Major General.

A man of great civic responsibility, as Republican Caucus Chairman and Appropriations Committee of the Washington State Legislature, he played a major role in the funding of the newly formed community college system, something in which he took enormous pride.

Robert authored a telling of his POW experience in a book, *Our Last Mission.* This kindly and easy-going gentleman died at the age of 96, January 25, 2014.

*What did your parents say when they found out you'd joined?*

Well, they weren't too happy. They had one son that was flying at the time and they didn't trust airplanes much. They though it was pretty dumb but they didn't say too much. They knew how upset I felt with getting fired from my very first job, especially since I anticipated having a long career in radio work. It didn't work out though, and Jean, my steady girlfriend at that time, in fact, she lived down in a little town of Thornton, about nine miles down below Rosalia. I was supposed to go down and have a date with her the night of the test, but they had dilated my eyes and said for me not to drive. In those days they were pretty careful about that, so I hitchhiked home again, from Gonzaga. I had to call her but I couldn't tell her why I couldn't come down to see her because my folks were standing right there and the phone was very public. As a result, she wouldn't speak to me for a few days until I could get her the explanation why I couldn't go down.

*It must have worked out, as you ultimately married.*

Yes, it's 61 years as of this year, so it worked. My folks totally accepted it all right. My brother was six months ahead of me in flying school, so when he graduated, he went to Langley Field in Virginia. When I graduated, I was kept right there as a flag instructor. So I stayed right there in San Antonio. I think my parents then were quite proud that they had two boys who were commissioned in the Army Air Corps and were both pilots.

*Did you and your brother communicate a lot while you were in the military? Was there any rivalry?*

Oh no. We use to write back and forth. He was about three-and-half years older than I was, and we didn't get along very well in our younger years. I don't know if you ever had a younger brother or older brother, but when I got in high school, he was just going into Washington State, and by the time I entered into Washington State he was out. We weren't very close at that time. Then, we both got in the service but strangely, there was no rivalry. He was ahead of me a little bit and he was a reserve officer when he went in because he had been in R.O.T.C., advanced military at Washington State. I did beat him in one promotion. From captain to major, and that got to him a little bit.

*What did he say?*

I was stationed in Maxwell Field, Alabama, at the time and he had been on sub-patrol down in the Caribbean, stationed at Puerto Rico. He was back then in Columbia, South Carolina, in a B-25 wing and I was flight instructor in B-24s at Maxwell Field in Montgomery, Alabama. So, when I got my major leaves and he was still a captain, I got a little old basic trainer that was on the field—an old BT-12 (it was a wooden airplane made by Beechcraft and they only made a few of them). I rattled in that thing all the way to South Carolina where my brother happened to be at the time. So when I got down there I went down to his squadron. He was landing and I got all his squadron mates out there on the ground. They all lined up, and I was having them pass in review and I was a major. My brother came in there and said, "What the heck is going on here with my troops?" He didn't know until he saw who I was. We had a lot of fun over that. But he beat me in the other promotions, so he got even.

*Is he still alive?*

Yes, he lives in Air Force Village in Riverside, California. He retired down there. He comes up occasionally. He's up here in the month of July, generally. So, we see him quite a bit then. We go to Maui in the wintertime, and he'll be over there at Maui, too. We've had some great trips and cruises together since we've both retired.

*What did he retire as?*

He retired a lieutenant general. He was stationed at the Pentagon and was on the air staff of the Joint Chiefs of Staff. When he got out, we had some really good trips to China, India, and Europe. We went to some of the capitals of Europe and cruised through the Greek Islands, so we've had a lot of fun with them. We still do.

*You got trained, you taught, and then what happened next?*

I stayed in the training command. I was in Texas when the war started. I was teaching advanced flying to students there. Then I went to B24s. We thought we were going overseas in 24s, but I was kept as an instructor again. It wasn't until the B29s came out that I kept volunteering for overseas assignments, which Jane hated with a passion. She couldn't understand why I would ever want to go overseas.

*What did you tell her?*

I just said that I was trained for that. This is the biggest show in the world, and I would have just felt cheated if I didn't get in it. It was a dumb attitude but I was twenty-two years old, and my good friends were all going to combat and coming back with BFSs and I never thought anything would ever happen to me. So, I kept on training. The B29s came out, and I had quite a bit of four-engine time by that time. I was picked to go to B29s at Hayes, Kansas. The wing I joined there was about halfway through their training. Though I didn't start out with them, I got in there as operating officer because I was a major and had a lot of four-engine time by that point. I joined that 73rd Saipan Bomb wing group, and we finished our training in Kansas and went to Saipan in the Northern Mariana Islands where we flew our missions.

Our first mission was at the submarine pins down there. That was kind of a warm-up practice mission. Then we started flying to Tokyo on a target we call the 357. We lost a lot of planes over the 357. There was an aircraft engine factory between Tokyo and Yokohama, and I was on my third run to 357 when we got hit. It ended the war for me.

*Tell me, did you parachute out?*

Yes. We were at 33,000 feet, leaving the formation, and Japanese fighters jumped us. Our intelligence said that the Japanese didn't have any fighters that would get up to 33,000 feet so that told us we were pretty safe, but they were up there and flying very well indeed.

*What were they?*

This was a Tony—a single engine that got us. It looked very much like our F51. We go him, too. The very first guy that came in on us got us. We saw him coming. We were on a nose, head-on pass, and we dropped our bombs and the bombardier has control of the top and bottom turret facing forward like that. The guy came in and he came out underneath us. I could see smoke coming out of his engine, so I was sure we had nailed him. I thought I saw fire flaming around him, but as it turned out, we did get him. By that time he had hit our gas tanks on my right wing, we were losing a lot of gas out of the right wing, probably about 500 gallons a minute! That's a lot of gas. I knew right then that we weren't going to go home. I called the engineer and had him start transferring gas into other tanks to see if we could get as far out as we could. We had three rescue subs between the Mariana Islands and Japan. There's Saipan, Guam, and Tinian. I had the radio guy start calling out, giving our position, and alerting the sub. So we could get as far away as we could, and ditch. Other planes had done this and some of them were successfully picked up. Then we caught on fire and there was just no chance after that. We were burning inside and out. Most of my controls were gone. My communication was gone with the guy in the back end.

*Did you get singed?*

Yes, I got burned when I went out of the airplane. It was not bad but was kind of a flash burn.

*How about your crew?*

As far as I could tell, nine got out. I had twelve aboard. I had an extra man flying with us out of wing headquarters—a full colonel. My ground commander was flying with me, instead of my regular co-pilot. I think my tail gunner was left in the plane. I'm sure my navigator was burned up because the fire started, as far as I could tell, around that area of the airplane.

*Do you remember their names?*

Oh sure. The navigator was Ben Edwards. He was a first lieutenant. My radioman was right in the same compartment and his name was Sgt. Wright.

*What do you remember about those guys? Do you remember any special stories?*

We trained together. The officers on the crew were pretty good friends. There is only one alive, now. He was the co-pilot that didn't fly with me that day. He lives in Texas. I've seen him a couple of times after the war. One of my gunners made it—Sgt. Schraeder. He was in solitary with me, and then he went out to the prison camp with me and he came home. The other members of the crew never made it. There were only three of us that survived out of the twelve.

*Do you remember a time when you knew you lost somebody in the camps?*

This Colonel Bruggy, out of wing headquarters, got out of the plane, and he was in a cell next to me. He died in the cell, though I was told later. I had no information on any other member of the crew. We had one other guy, Sgt. Golfrey, one of the gunners, that was in the staples—in the cell. I think that it was Christmas Day that they came in and took him out. He had to walk in front of my cell to get to the exit. He was fine as far as I could tell. He had a flying jacket on, and he was tied up and blindfolded when they took him out. Never a trace came of him later. After the war, of course, we tried to find out all members of the crew—and there just was no trace whatsoever.

The Japanese, many years later, put out a report that some of them had died from gunshot wounds. My Lieutenant Ward, my engineer died from a gunshot wound. You just never know if they were executed or killed. When I got on the ground—I was cornered. It took about half an hour to get down. When you bale out at 33,000 feet over Tokyo, there are an awful lot of people who see you come down. I thought at first, "What do I do?" I was scared and didn't know what to do! I thought I'd sneak up into the mountains, get in a cave, and stay the war out. That wasn't very practical. I wasn't going to make it in a cave during the cold Japanese winter. Then I had an idea that I'd sneak down to the bay—to the water and see if I couldn't steal a boat and get away from there. I was six-foot one, and in flying clothes. So I couldn't sneak any place in Japan. It didn't take them long to corner me. They tied me up with ropes and let the civilians have a go for a while and they gave me a real good beating. I found out afterwards, fifty years later, when I went back to Japan and met with some people, that when the civilians intended on killing me, (I was actually in a little clearing there in a school yard as my plane had crashed there), but there was some village leader there that came out and said, "No we are not going to kill him." So he saved my life. Fifty some years later, when I went back to that area, I met this fellow. He presented this paper to me. This is a translation. "A pilot Robert was on board an airplane coming with a fire. At the moment just before it came apart, he was seen in the sky. After 50 years, he goes round and round the battle. At the time, he was not invited but was ardently welcome today."

*That is great—a piece of history. Tell me about this man. That must have been pretty heart rending to meet this guy.*

Yes. I hadn't realized at that time that he had saved my life. I knew they were pounding on me, but I didn't know they were intending on killing me there. Although they did kill some people. At any rate, he saved me. When my wife and I went back to Japan four years ago, they had arranged a "going in peace" ceremony back there. Though a friend of ours —a Japanese lady we met on Maui. She was kind of a historian. She made a big investigation here. It took her about five years, but she found out where my plane went down. She tracked down people who saw my parachute come down and found the guy that was in on my first capture, and also the fellow who saved me.

*How old a man?*

He was 90 when we met. He'd be 94 now. I think he is still alive. I have not heard that he isn't. He couldn't speak any English but he was all stooped over. He had a cane and was a small little tiny, Japanese fellow, but he had been a leader in the community there. I was quite touched to meet him. We had our pictures taken together. We went right back to the place where I landed my parachute. There's a golf course there now.

*Did you shed some tears?*

Well a little bit. I looked over that area. I was telling Jean that, when I got out of my parachute, I wasn't thinking too clearly. My hands and arms were all skinned up. The parachute contained a little survival kit, which had morphine, sulfur powder and all that. I should have just sprinkled that on me. But I didn't and I had a lot of trouble later on with infection in my hands. I just walked off and left all that stuff. I walked down this road that slowly curved to the right. On top of a bank there was a little Japanese lady standing there. I didn't know what to do. I just kind of smiled at her, and she just scurried off and went into a house. I walked around that road. When I was there four years ago, I described that and sure enough, they were able to take us back to that road. So 50 years later, I walked down that road with Jean. Then I had some tears thinking of the time that had gone by, and how 50 years ago I had walked down that road in bad shape.

*What was the hardest adjustment to the prison life?*

I suppose that the hardest thing was the hunger. You never get adjusted to that. It was a very cold winter and those cells were eight by ten feet— absolutely bare—just a hole in the floor for a latrine. We'd use four little Japanese blankets at night. It was bitterly cold with no heat, at all. After the hunger sets in, that sort of rules your days and nights. I got down to about 85 lbs. from about 175 lbs. It was a starvation diet. I don't think, if they hadn't taken us out to the prison camp, where we got a little more food, we would have ever survived those cells. I would have not made it through another winter—I know that. I was in prison for 10 months.

# Many Missions

## Robert J. Carlson

Robert J. Carlson graduated from Rogers High School in Spokane and went in to the service early in 1942. He applied for aviation cadets, went through and made it, all the while wanting to be a fighter pilot. After qualifying as one, he flew a P-39 for fighter training in Thomasville, Georgia, in 1943. He further trained as a tactical reconnaissance pilot in the brand-new P-51B in Meridian, Mississippi. Overseas, he went and to Naples, Italy, where he started flying missions, P- 51s for tactical reconnaissance. This was when they were first starting flying their missions back in to enemy territory. He was always close to the battle lines. He flew that, and then on up almost to Germany. Robert had 82 combat missions.

Later, Robert went to Georgia Tech and got his degree in engineering and then was recalled to service in 1952. He'd been working for Ford Motor Company as an engineer for about two years, and then went back into combat. He was sent to Korea and to forward air control type of work. There he flew combat missions. He finished that up after 20 combat missions and nine months on the ground as a forward air controller.

Then Robert returned to the states, to the maintenance business. He served also as a flight instructor, and later as test pilot for various types of aircraft. including helicopters. Then he ended up with a couple of headquarters jobs. Finally, he returned to Korea again to start an airborne radio relay.

*So you were in the training command.*

I spent several years in training command and was instrumental in putting our people in to the training of people for F-4s. Our various fighter outfits were coming in to the new F-4s. Prior to that, I had the training for the B-58s in the late '50's to the early '60's. From '59 until '62, I had a field training detachment in Texas. That was on the B-58, with B-58 maintenance and field training. I ended up in Texas in 1971, where I retired.

*Let's talk about your training when you went in the military*

.

Well, it's hard for me to put into words. Probably the strangest thing occurred because I was a Harley-Davidson rider back in those days, and I had a slight accident. I had a broken leg when I got my call to go to Cadets in 1942. So, I thought, "Well, I'm going to go anyhow." I lived in Spokane at the time, and my call would take me to Nashville Tennessee, to begin my training. Well, I didn't know anything about the service at that time, but I thought, "Heck, I'll go for a good train ride anyhow." I went, crutches, cast, and all, and turned myself in at Nashville, Tennessee. When I got off the bus, the people didn't know what to do with me. Here I was on crutches with a cast on the leg clear up to the hip. I said, "Well, if you don't know, send me down to the hospital. I'll be all right there. It'll only take me about a month and I'll be ready to go again." Motorcyclists are used to broken bones anyhow. So, that's what happened. Surprisingly, I met the hospital commander and told him, "Well, just put me in here for a few weeks and this thing will be all right. I'll be okay," and darned if they didn't do it. I was kind of a strange one.

*Did the military wash out any friends of yours during training?*

I can't recall on them. I figured I'd be one of them because I ran into problems. When I was getting ready to solo, I had no experience whatsoever, only about eight or ten hours. I was ready to solo, but I could not get that PT20 to land. I knew I wanted to do it, but it would bounce, and I'd have to correct for it. Of course, back in those days, all you had was one-way communication. The instructor could yell in to this phone to some earphones that you had on. So, all you had to go by was what he told you to do. You'd better, by golly, do it. He really got on my case. I thought, "Oh boy, I can't do this. Darn thing. I can't do this right. Why don't we just go home and quit for a while?" I thought that was it, that I was through. I flew on home, landed and got out, and the instructor said, "Well, as soon as you cool down, I'll see you tomorrow." I felt really dejected because I figured I'd blown it. The next morning, he said, "Well, how do you feel?" I said, "Well, kind of foolish." He said, "You want to try it again?" I said, "Heck yes, if I've got a chance." I didn't think I'd have another chance. He said, "Shoot. Don't worry about that. It takes a long time to learn to do these things." That was a good lesson. I kept working at it and managed to get through there fine. Then I went through Basic in old BT-13s. That went quite well, but there are always things that you don't catch onto quickly enough. I kept thinking, "Man, I'm never going to make it here." All these guys I roomed with were all completely confident. I guess I was the only one who wasn't. Lo and behold, those three guys all washed out. I was the only one who made it through Basic. By now, my confidence was getting a lot better. Then came advanced flying school at Luke Field in Phoenix, Arizona. There were T-6s on that base, for advanced trainer. There were a few people whom they picked to get 10 hours in a P-40. In August of 1943, I was picked to get my 10 hours in P-40s. That was really flying high then. I was still hoping to make fighters, to get into fighters. The optimum would be the P-51s at that time, but we flew 40's. I guess it was about October by the time I got to Thomasville, Georgia. That was with P39s.

*Was the P-39 tough?*

No, I didn't think so. You had to be pretty careful with it. You have to really learn to fly that thing, but I'd also been keeping my eyes open for possible flight test jobs whenever they might come up. I didn't get that until a year or so later. The P-39 was a lot of fun to fly, but you had to be careful with it on stalls. I lost several good friends. One died on the final, believe it or not.

*Tell me about your friends.*

They were just regular guys. They made it all the way through flying school and that. We all knew each other pretty well. One time in particular, I remember, that we had gunnery practice out in to the Gulf. We had a fairly high altitude for those P-39s. They didn't fly well at high altitude at all. Then we'd come back, and we'd usually fly in a four-ship formation. I'm flying Tail End Charlie, which was number four. We were up in these pockets under these nice, white fluffy clouds. Well, that's always a lot of fun, if they don't lead you into them. If you're flying number four, it's tough enough to stay up with everyone. Anyway, this particular flight leader led me into the clouds. Here I am, almost colliding with number three, and it scared the heck out of me, so I just flat broke off, which you're not supposed to do in training. I went home and landed that P-39 and went stomping on into the place. Nervously, I waited for the rest of them to come back, figuring I'd blown it again. I figured now that I was going to have to go fly bombers or something like that. Or worse. Fortunately, the flight leader wasn't too concerned. I did have to see the flight commander. He didn't like what I did at all. He said, "Well, you are set to climb the tower steps on your weekend few days off, on a Sunday." So I ended up having to spend about three hours climbing up and down those steps for that breach of discipline. I said, "I'm not flying with that guy again. Anybody who will take you into clouds, not watching what he's doing, I'm not going to fly with him again." Of course, you just don't talk like that when you're a Second Lieutenant, and telling some Captain what you're going to do and what you're not going to do. But I got away with it.

I climbed the steps that particular Sunday. He came out and said, "How are you doing Colonel?" I said, "Well, I'm still kind of hot under the collar about all of this I'm doing on my day off. Don't do me any favors." Now, you've got to remember this as back in the days when you had gasoline rations, and sugar rationing, and everything else. None of us had cars, but he did. We couldn't afford them. He said, "I want to go down and see my friends Tampa. I need to go down and have someone take me to catch a plane, the C-47. You can have my car. I've got plenty of gas in it. Why not take it for the weekend?" I grudgingly agreed and, as it turned out, another friend had set up a blind date for me. After all, I had a car for a change. That blind date ended up as my wife. That was 58 years ago. We've been married 58 years. That was the first blind date I had with her and it was with that car. The funny part of it was that she had been going with the leader, who dictated I climb the steps. That's the strange part about it.

*So you got even with him, after all, didn't you?*

Yeah, but Russell became a good friend after that, anyhow.

*What struck you about this lady? Obviously, she made a great impression, didn't she?*

She made more of an impression on me than I did on her. I made a very bad impression on her. It so happened that she'd been going with Russell, but I didn't know this. I guess the two had some phonograph records in back. One time four of us went to one of those neighboring towns to have a dinner and a little dancing. Something came up about some records in the back seat. I said, "Don't worry about them. Sit on them and break them. He doesn't care." Little did I realize some of them were hers. At the time, nothing was said. Nothing was said at all, but I didn't hear about that until way afterwards. Anyway, we developed a good friendship. So, time comes to leave, and I didn't know if I was going to head for combat at that time or not, but they did send me to this other training camp in Meridian, Mississippi, which, when we got there, was brand new P-51Bs, which really made me happy. I went through that particular training and then on to combat with the *General Anderson*, which was a new troopship heading across.

Here I am, a proud new Second Lieutenant, and I get the job of being a KP pusher. They had 5,000 troops on that thing, so that KP-pusher job went from early in the morning to late in the afternoon. We fed people two times a day, starting at about six in the morning until about ten, and then from about two until about six at night. Most of the troops were sick, at least half of them were, and there was no place to sit down. That ship, which was a new transport, would roll. It didn't have quite enough balance in the bottom. It would roll up to about 45 degrees as we were going across. That was in the days when the German U-boats were quite prevalent.

*Did you get sick on board?*

Oh, Lord yeah. I'll tell you, it's hard to figure you'd even stay on the job. I don't think I could have eaten anything and stayed on the job. If you didn't bother to eat anything, that'd be all right. That was kind of a tough job to take for a Second Lieutenant, being down there pushing the people through food lines. It wasn't a fun job for a young fighter pilot, ready to go.

*Let's talk about some of your more interesting missions. What was the most interesting mission you went on?*

There were some fairly interesting missions a couple of weeks after we got there. There was a big push, and the officers decided that we could fly off of a little strip. It was a pretty anxious time because the Germans could fire 88's at you. You had to get up, get out of there, get out to sea, and when you came back in, you'd stay pretty low until you knew right where you were. There was a sunken PBY out in the bay, so we knew right where to come and get our wheels down. We had to land as fast as we could before they had time to fire at us. That went on for maybe a week or two.

We flew some missions in to Northern Italy, strictly looking for what was happening. That was back when the Germans had one of those big cannons that they had mounted on railway cars. They'd pull them in to a tunnel and leave them in there during the daytime. Then they'd pull them out and fire at night. You had to watch for this, so we'd do some night missions, but not much because we didn't have any lights or anything. We had just plain old runways. No control towers, so it got pretty interesting at times.

*Was there a time that you really took some shot?*

No, not too badly. I've been fired on quite a lot from small arms fire, from down low. We flew fairly low. We usually flew at six or seven thousand feet, so we had to be pretty careful when we were in these areas. We'd find trains backed up in to enemy territory, and we'd lead the fighters in to them. Sometimes we'd shoot up the engine, if we thought we could get away with it.

*Describe shooting up the engine, and describe one of those situations.*

We would make an attempt at perhaps shooting the engine on a train to help slow it down a little bit. Then we'd get out of there and leave. I managed to get one victory. I shot down an JU88 one morning. It was back about 200 miles behind the enemy lines, and we were able to intercept. I was flying wing, and, not long after I got there, my leader had gone in, but his guns jammed on him. We were coming in pretty fast, so I got in there behind him. I almost overran him, but not quite. Next, I got in firing position behind him and I managed to hit the enemy. He went into the Mediterranean. We were pretty low down to the hills when it happened.

*How low were you do you suppose?*

Oh, we were probably about 6,000 feet when we saw him, and we could see what it was. It was a twin-engine. It was fast. It was damn fast, nearly as fast as we were, but we got the jump on him, and we went screaming down there. It's one of those things that you need practice for, but we hadn't had any practice on that. Anyway, I lucked out. I shot him down and got credit for him.

*So you were pretty sky high after that, weren't you?*

Oh yeah. You feel pretty good. You pat yourself on the back several times on something like that. Just before the Southern France invasion, we were flying off of Corsica, across the Mediterranean to Southern France, and then up the river almost as far as Nantes. We were a little bit skeptical, because there weren't many fighters up at that time, so we didn't have too much problem from air-to-air, but, every once in a while, we had to fly with wing tanks in P51s. Of course, you don't want to get in to scrap with wing tanks on the P51, so you'd have to drop your wing tanks. Well, that presented a problem. You couldn't go after these German training planes, because, if you dropped your wing tanks, you wouldn't make it back to Corsica. You'd end up being at the mercy of the Germans, if you had to bail out. So, we had to pass up on a few good opportunities.

*What's the biggest raid you went on?*

When German troops had been stalled all winter down around on a line north of Naples, that was probably was the biggest, and most fun, of any of the flights that I had. It was great getting them on the run, finding tanks and even calling in the shots from the Navy. We'd adjust fire from the Navy ships out in the Mediterranean.

*Is there any image that sticks in your mind?*

They still had some horse-drawn artillery, believe it or not. They would get out in the daytime with those things. I've gone down to slow them down and shoot them up a bit, then would climb up and call in the fighters. You've got to be pretty careful, because you're pretty much down in the mountains, and you're pretty much a target yourself, if you're not careful.

*Did you lose any friends that way?*

Yeah. There was a truck one time, and I was leading. My shots apparently blew up the truck, but I'd already pulled up. I hollered at my wing man to pull up. He didn't hear me and kept on going. Anyway, he went through the concussion and lost his P51 wing. It came off, and, of course, he went in, but that was the only personal one I happened to be on where I lost a friend.

Another time, my best friend was shot down. He was captured and walking, but he didn't pay too much attention to what the Germans were telling him to do, so they shot him and killed him. We found out from a British soldier with whom he had teamed. The British soldier had stopped and done what the Germans told him to do, but he didn't stop. He said, "Go to Hell" and kept on going. The Germans wouldn't take it. They shot and killed him. I had other friends who were lost in France on that day, for crying out loud. They were flying P-47s. One was my instructor. He was killed in his P-47, when taxiing out to take off. It turned out a B-17 unloaded on him in the wrong place. He got three or four of our fighters. That actually happened. Of course, that was a distance away from us. We just heard about that. Our 51s were pretty susceptible to ground fire, because we had that scoop that went back and was under the belly. You had to have time to get enough altitude to bail out safely before that engine seized up. Of course, you had to get out of it. There was no bringing it down, especially over water. Most of our guys got back that had to bail out.

*Was that a pretty rough thing when you lost your friend?*

Your mind becomes pretty steeled for it, by the time you get in to combat. You know it's going to happen. You realize that he was doing his job the way he thought he ought to do it. It happened to him. Well, it happens. You might be next, but you never stop. I'm amazed. Today, I watch these people in tanks where they've got the reporters embedded over there in Iraq. I'm amazed at how they keep going on, but I understand it. It's perfectly understandable how they keep going on. You just do. Really, you don't even consider any other choice. You just figure, "Well, I'm going to make it through here as best I can." You don't ever figure you're going to get clobbered.

*Was there one day of flying that sort of stands out?*

Golly, I've been an instructor pilot for a number of years, both in propeller jobs and then jets. I got checked out in jets fairly early, when they first started having them in 1954, I was an instructor pilot in T-33s for several years, and I always enjoyed that. I enjoyed it a lot. Then I was a quality control officer at a base where we had the only helicopter flight school. They had a helicopter flight school, and Air Force at San Marquis, Texas. We also had started to train Army pilots, so my particular job, since I was an instructor pilot in T-33s, was a job in quality control where I could fly some real good sergeants. I could fly as an instructor pilot and a test pilot most of the time. I loved that job. It was the one I enjoyed most of all. I enjoyed flying test flights and various types of airplanes, and helicopters.

*Tell me about any unusual things that happened with helicopters.*

Well, compared to what we've got now, they were very primitive. There were H-19s, and H-21s. which the Air Force had. The H-21 was a twin-rotor, which I didn't care much about, but the H-13s were pretty interesting.

*Any close calls in helicopters?*

I had a helicopter crash when it quit on me. If your engine quits at a low altitude, you're going to come down in a chunk. That's what happened to me on a test flight. The engine quit out over a railroad cut and a bunch of mesquite trees. I didn't have much speed and didn't have any altitude, 30 or 40 feet. I came down as best as I could. I tried to give it as much pitch but it was "Whoop, Whoop, Whoop, and Blam!" Fortunately, I was all right. My crew chief was flying with me, because we needed to check these things out. Of course, the rudder, the big blades came down when we hit and whacked off the tail of the helicopter. The foot control, which controls the pitch of the tail rotor, is on cables, and my crew chief had his feet on the floor. The pedal is just a little pipe on the end of a foot pedal, so it came back up and broke his leg. I had set all the switches off when I knew we were going to hit, so that we wouldn't burn. We landed on a cactus and ended up upside down in one of those plastic-bubbled canopies. I managed to get him out. I wasn't hurt. Unfortunately, nobody had seen us. Of course, we weren't in radio contact and were out in the test flight area about a half-mile to get out where you could hitchhike a ride to the hospital. When I reached the hospital, I yelled, "I've got to get a flight surgeon and an ambulance fast." I started walking toward the flight surgeon's office and then I spotted him running down the hall, He was a friend of mine. He said, "I haven't got time to talk to you now. We've got a crash out here." (Laughs) I said, "Wait a damn minute. I'm in that damn crash. We've got to get out there. I'm the only one who knows where my crew chief is. He's out in the darn mesquite somewhere. He's about a mile in to this stuff." He'd been running down the hall to go get an ambulance and didn't know where to go.

*Was there some big discovery or cautionary note that you recall?*

I was also an instructor pilot for T-28s and B-25s. I was flying one afternoon to evening, with the Club Officer from San Marquis, and, of course, I was used to the T-28. It's pretty severe when you practice a spin. We were practicing some stalls and spins, but I'd never been up with this particular man before. Suddenly, he froze on the controls on this spin. I couldn't get him loose from it. I yelled at him. I finally said, "Another 2000 feet and you'll be on your own. I'm bailing out of this thing, if I can't get you loose." He finally paid attention. Of course, I still had room enough room below, so I was able to get us out of trouble, but that was the closest I ever came to bailing out.

*It sounds as if you are a cool customer.*

Well, I'm thinking ahead, but I'd never had anyone freeze on the controls in a condition like that. That's a critical place to have him freeze on controls. Needless to say, he was not passed and was later grounded. It doesn't happen often, but it does happen. It happened to me. I've had things happen to me that scared the heck out of me, too. I've had P-40 engines quit when they began to ice up. You begin to wonder just what you are going to do, so I put the carburetor heat on it, which I should have already done.

The P-51 was the most fun to fly and then it was the T-33. I enjoyed the T-33 a lot. In fact, I enjoyed instructing in the T-33. I've flown C-54s. The B-25 would come third. I like the B-25. As to why I liked it, I just liked it. It was a good handling airplane, and it was a fun airplane to fly. I never had to fly it in combat or anything. It was always for support function. I always prided myself on being able to keep up with the different types of airplanes, and I'd fly them around the same time. In base flight, when I had that job, I had two C-45s, one C-47, and two or three T-28s. For a while I had about four T-33s. Besides that, flying helicopters, and also L-19s and L-21s, when we were training the Army pilots. This is back in 1954, 1955.

*It sounds like you were a daredevil, riding motorcycles around. As a kid, were you a daredevil?*

Oh, I'm afraid so. Yes, I had a big old Harley that I really, really enjoyed. At the time, I had an uncle and aunt in Seattle. I'd ride over there on a Friday night and come back on a Sunday afternoon. I'd just love to go, and I really enjoyed that because we had no interstate. You had to be pretty careful. I was coming back one Sunday afternoon, and I was going through Wenatchee, Washington. I was coming up through those apple orchards along the Columbia River before it climbed up out of the Gorge. I was really rolling, when an old Chevrolet came toward me. I thought, "Uh Oh." I didn't know whether this guy was going to stop or not. I was about in the middle but was going too fast. I got over, until I couldn't get over any farther. I tried, but I couldn't miss him. I hit his car on the left front and that was all that I remembered for a few days. A young couple that I'd just passed moments before on the road, and had given a salute to, stopped and picked me up. This guy who hit me was an itinerant fruit picker. When I saw him coming, I probably slowed to 70 mph. With a motorcycle, at that speed, if you try to move over, you can't hardly do it. Of course, I had no business riding at that speed.

*How long did it take for you to get rid of that wild and crazy streak?*

I guess that lasted through flying the fighters in combat. I'm pretty sure of that. I still had pretty much of it in 1952, and when I was recalled in 1953 to Korea. If I'd had a good airplane, like the P51, I'd have liked it better. I volunteered for F-86. Of course, I had no jet time at that time, so I went down, volunteered, and you could hear the laughter all over the tents. "Who do you think you are? We've got lines of people waiting to try to fly these F-86s."

*Did you have any real classic flights that you remember from Korea?*

No, there were pretty common type things. Your job as Forward Air Control was to pick out areas, talk to your ground operators, and see what problems they were having. In the early part of '53, they had quit fighting in July sometime. All we'd do is bring the fighters in and put them in on a suspected tunnel. Dugouts, tunnels were abundant on the other side of the lines. We would make sure they didn't come down on our side, which we realized in WWII was a big problem. That's why the tactical control groups were started in Korea. So, that was our primary job.

*Tell me a story about your first experience with flying jets, and training.*

It was great. Of course, I was really perturbed when I was over there in Korea and wasn't able to get to go where I wanted to go because I didn't have any jet time. As soon as I got back in a state- side assignment, I put in for jet flying upgrading, and I got it. I went to Selma, Alabama. I loved every bit of it. I wasn't taken aback at all. I enjoyed it. I didn't have any close calls. The only thing that I remember that upset me was flying in close formation at night, which we hadn't learned to do. I hadn't done any formation flying at night. One time I got vertigo, but they'll always tell you, "Be careful. Use your instruments." You're having to watch the other airplane, but use your instruments and stay with anyone. If another goes upside down, you go upside down. That always saved me. We could talk to the other aircraft by that time, usually. You'd tell them what had happened, and say, "I want to break away. I'm going to go straight and level here for a little bit."

*Did you ever hot dog, buzz anybody, or do any crazy stuff like that?*

Yeah. I was down under the bridge up there in Mississippi. I can't even think of the name of it right now, but it's on I-20 now. It used to be old US 80. That one was a scary one, because I knew I had plenty of altitude underneath it, but I shouldn't have been doing it. I just split from about 12,000 feet and went down to about 500. I went under that bridge, and about a second before I got under it, I realized that there were some power lines there that I hadn't planned on. I hadn't been down and looked, and I had to, really quickly, go forward on the stick. That was a scary one. Stupid. Completely stupid.

*You obviously gained some good sense in life because you made it through, right?*

Oh yeah. I made it through.

*It's got to be a theme that you were enough under control. You were under control when you had to be. How many hours did you fly?*

Oh, I've got about 5,000 hours. I didn't fly any after I retired. I would have liked to have, but I just wasn't in the position, and it wasn't affordable. I had a survey and engineering business. I thought about getting a small airplane, like a Cessna 172, to fly places, but it didn't have enough real need. I could have written it off. I always thought that I should have.

*So, what did you think when you were out there surveying the land, and you saw a plane pass? You never thought, "Oh, I should be up there."*

Well, to some extent. That lasted for a while, but it finally gets to the point where you don't think too much about it. I still would probably enjoy flying. I just didn't do much of it after I retired. I'd gotten pretty tired of flying those big old four-engine jobs. I didn't care much about that. That was strictly boring, except when something went wrong. It was a lot safer than the old single-engines. When something went wrong, you had to act quickly. With four engines, you just shut that one down and go on like you were going before.

*Did you ever have to lose a couple of engines on the four-engine?*

No. One engine at a time. Always lucky. Once I lost a prop problem at St. Louis around a bunch of trainers. I had just lifted the nose off, when the crew chief hollered, "Runaway Prop!" I slapped the nose wheel down and pulled all the throttles back. We were able to stop. We were lucky. I didn't know whether we could or not, but we had no other choice. We'd have probably gone in the Mississippi river.

At other times, when getting set up for our airborne radio relay in 1970, we had to go over to Japan. We had to find our airplanes and then get ourselves checked out again. Get the airplanes going. We spent weeks there. When we'd finally get one of our airplanes, we'd need to get all four engines to run at the same time long enough to take off. Something would always happen to one or two of them while on the runaway, so we'd pull the plane off the runway, taxi back, and they'd work on them for a while, and then we'd try them again. We had quite a time. This was 1970, and we had airplanes that were really worn out from the Berlin Airlift, back in '47. Some of them had as much as 50,000 hours on them, and you've got to be careful with airplanes that had been flying for that long.

*Were you frightened by that?*

Not really frightened, just aware, very much aware of anything that's going on. It sure sharpened up your senses. We weren't considered test pilots, but we were, those of us that were flying the things.

*You just don't like them when they're worn out.*

No, no. The DC-7 or the DC-6 would have been a lot nicer. It's kind of like going from a Model T to an early V-8 or something like that.

*How many planes did you check out on?*

Usually not more than one or two at a time, but I probably had 30, maybe.

*Well, it's been a great life of flight, though, hasn't it?*

I wouldn't trade it for anything, Irv. I wouldn't trade it for anything. The only thing I might do different is try harder to get the exact jobs I wanted. I guess I was pretty lucky all the way through. I didn't try that hard, other than just trying to do everything right, or not get in too much trouble. I got in my share of trouble all right. I always managed to wiggle out of it some way or another. I made Lieutenant Colonel in 1966. I probably could have made full colonel. It's my own fault that I didn't.

*How do you feel when you see these pilots today in the war on television?*

I have to envy them. I envy them and what they're doing. At the same time, I'm a little bit followed how have to keep up with radars, and everything like that. These heat-seeking missiles that they have nowadays just wouldn't make me comfortable up there. We had countermeasures. It would unsettle me, but I have great respect for them. Believe me.

*No different than riding a Harley at 90 mile-per-hour, or doing some of the flying you did.*

Back in those days probably. It's probably safer than that!

# Over The Hump

## Roy S. Waldau

Roy S. Waldau is a most colorful and fascinating person, with an extraordinary range of life experience. He served almost two years of war-time service in Combat Cargo, where he flew nearly two-thousand hours and made sixty trips over the Hump as part of the China-Burma-India (CBI) Theater of Operations, transporting critical elements. Once retired, this fascinating, pipe-smoking original continued lead a colorful life as a Pinkerton detective, professor and published writer of note, which included his classic *Vintage Years of the Theatre Guild, 1928-1939*. This American original died in 2008.

*Roy, it was a little unusual, your rank. You flew as a staff-sergeant. How'd that happen?*

Yes, here's what happened. I made application as an air cadet. Orders then directed me to Georgia Tech. "If you completed the training, would you be willing to accept a commission in the Army Air Corps," I was asked. It was proforma, but then the guy said, "If you become commissioned, you'll have to agree to serve in the Air National Guard after the war is over." I answered "no," so that must have been the reason.

*What did you transport?*

One time we flew Prince Aly Khan, who later married Rita Hayworth. We flew supplies, mules, the usual stuff. One of the things was aviation gasoline, which was highly flammable. We carried a plane load of twenty barrels and always tried to protect it by placing the barrels in the middle of the plane and buffering around the sides, so if bullets came in, they might not puncture the cans. When we got to Kunming, we unloaded it. The Chinese would then hoard it. One time, when we flew in, we were really low on fuel. The Chinese didn't want to give us fuel. We had to negotiate with them for fifty gallons to get back to base. The gas was located up behind the engine. The Chinese commander ordered a guy to get up on the wing and replenish the gas, but the guy didn't want to do it. He said, "No, no." The commander shot him dead. It was brutal. Another Chinese climbed up and put gas in the plane.

*You got shot down once on a mission and parachuted out.*

We were supposed to dump moldy, magot-infested rice. This was near Pym in what was Burma at the time. I was co-pilot on this mission. This officer had this idea that our plane would fly in first and drop this rice behind Japanese lines. The Japanese, he figured, were starving and would run to the rice. That would take attention away from shooting at our planes. Unfortunately, our plane got hit and it went down. Every one of us hit the silk, but the crew chief, and he went down with the plane and was killed. I landed in my parachute in a plowed field and busted up my knee. I dragged myself, maybe fifty yards, to a group of people standing around. The villagers there were lovely people. They took me in and sent for a jeep, which picked me up and got me back to the British office. These Burmese people were lovely people.

*Any embarrassing moments?*

We were flying Red Cross people and were the first Americans to land at Tientsin. To land there, you are looking at a hollow, which was like a dish. You'd roll down one way and have to roll up the other side. In this, you could land from any point of the compass. With most landing fields, the runways line up according to the prevailing winds. We flew in and there was nobody in the tower, so the next thing you look for is a windsock to determine wind direction. Lacking those, you drop a smoke bomb to see about the winds, but we had no smoke bombs. Sure enough, about a mile away, there was a stream of smoke. Based on that, I began the descent.

When I saw the landscape ahead, I tried to put on the brake and the tail came up, the landing space was fast coming up. When I hit the runway, I realized I was downwind. We landed off the end of this dish. It was a swamp and the plane came to rest on its belly. It was all mush and mud there. My face was red with embarrassment. I said, "How could I have misread smoke like that?" I walked over and it wasn't smoke at all. It was laundry strung out on bushes. Now it was my responsibility to get us off the runway, so we hitched up this water buffalo to pull us out of the mud and get us back on.

On one airfield, we were told to be careful. We landed there in this red mud. It was monsoon season, so the water would collect in these holes. We knew it was going to be hard to take off. We didn't know what to do. Should we stay overnight? That was met with "If you want to meet the Japanese." It turned out the British controlled the field during the day, but the Japanese took over at night, where they used it to drop supplies. We heard that and the pilot said, "The hell with that. Let's get out of here," so we had four guys lift the tail off the ground, to help us get clear.

On one occasion, we flew around the rim of North Africa to get over to Alexandria and Abadan. We flew over the Taj Mahal. It was night and it was lighted up, which I thought was unusual. It was beautiful and we circled it.

I can remember walking out into the rain and seeing these religious scrolls that had been thrown out there and were rotting. I gathered them up, as many as I could, and I put them underneath a shelter. They were Buddhist texts.

*You flew the Gurkhas. Ever intimidated by them? They were fierce fighters.*

No, they were very sweet people.

In India, at the end of the war, the Indians wanted their independence and they wanted it now. I was riding in a ricksha and had a Gurkha hat, sort of like an Australian hat, that had been issued. This guy got on the back of the ricksha and hit me on the top of the head.

*What did you do?*

I was stunned and staggered out of the ricksha and made my way to another conveyance. There was this negative feeling toward the British among some after the war. The only way I could be identified me was that hat, which helped some to cushion the blow.

*Talk about any characters you knew?*

With us was a boy named Taboo, a little Chinese orphan whom the squad had rescued in China. He was something; he had a vocabulary of dirty words worse than ours. Along the Bay of Bengal, there are beaches that are black, made up of lava. You could actually land on the beach, if you didn't have a load. The pilot said, "Let's go on in." Taboo wanted to go swimming. On the beach, we noticed these transparent clumps all over the beach, dried out, cooked by the sun. We didn't know what they were at the time, but they were Portuguese man-of-war, and they were lurking everywhere in the water. I went in and got stung in the crotch. Did that get me out of that water fast? But poor Taboo got it worse. He was stung all over his body and was really suffering. He had trouble breathing, but he survived.

*Another dramatic story?*

When I went over as a replacement pilot, the squadron had been in China and withdrew to Calcutta. That's where I joined them. I'd sometimes fly with Colonel Robert Cowan, an old Flying Tiger and formerly an airmail pilot, who now was flying a P-40. He really had a bug up his ass. Cowan insisted we fly in tight formation.

Years earlier, the *Ladies Home Journal* magazine had a series by Monro Leaf called, "Is There a Watchbird Watching You?" It would include things like "Brush your teeth, the Watchbird is watching you." Anyway, the colonel had the Watchbird painted on his plane. Sometimes, we would fly with him for three or four nerve-racking hours, in close formation, which was really scary. After awhile, we'd try to spread out, but whenever that happened, Cowan, the Watchbird, would race up, and we'd fly back into tight formation again (Laughs).

One of the original guys in China was a guy known as "Pinky." I was co-pilot with him. A C-47 doesn't fly fast, usually around 120 mph. That's good, which helps in controlling. I liked that, but Pinky liked to push the limits. He'd pull buzz jobs on railroad trains. We'd go down so low you could see the faces of the people. Once, flying low, I looked up and there was a semaphore with its poles and flags. I pulled us up right away and we just missed it. Pinky had always seemed to try to look older. He wore a mustache and a beard. One time, we had birthday party for him, and it turned out it was his eighteenth birthday. He was happy because they couldn't kick him out of the Air Force anymore.

*Did you lose any close friends over there?*

No, close friends, I can't say so. I did have a close friend that I wish I could have kept up with. He was a pilot and flew in another outfit. We liked the same things. We laughed at the same kind of jokes. We thought exactly alike. We were both interested in theater; we both read the same things. He was a Thomas Mann nut. I liked Thomas Mann. If I had a brother, I don't think I could have been closer to that brother. He'd come over and have a beer. I'd go over to see him and we'd drink Canadian whiskey someone sent. I was originally from Connecticut. He had lived in Calgary all his life. We tried to figure out why we spoke with the same accent. We figured we'd never been closer than 3,000 miles in our life, yet there was a closeness: this comrade in arms. This is probably the only place on earth where this would happen.

Made in United States
Orlando, FL
11 November 2026

90609408R00168